Sacred Bovines

Sacred Bovines

THE IRONIES OF MISPLACED ASSUMPTIONS IN BIOLOGY

Douglas Allchin

OXFORD
UNIVERSITY PRESS

OXFORD
UNIVERSITY PRESS

Oxford University Press is a department of the University of Oxford. It furthers
the University's objective of excellence in research, scholarship, and education
by publishing worldwide. Oxford is a registered trade mark of Oxford University
Press in the UK and certain other countries.

Published in the United States of America by Oxford University Press
198 Madison Avenue, New York, NY 10016, United States of America.

Library of Congress Cataloging-in-Publication Data
Names: Allchin, Douglas, 1956– author.
Title: Sacred bovines : the ironies of misplaced assumptions in biology /
Douglas Allchin.
Description: New York : Oxford University Press, 2017. |
Includes bibliographical references and index.
Identifiers: LCCN 2016050784 | ISBN 9780190490362 (Hardback)
Subjects: LCSH: Biology—Research—Methodology. | Biology—Study and
teaching. | Fallacies (Logic) | Errors, Scientific.
Classification: LCC QH324.A45 2017 | DDC 570.72—dc23
LC record available at https://lccn.loc.gov/2016050784

1 3 5 7 9 8 6 4 2

Printed by Sheridan Books, Inc., United States of America

CONTENTS

PART V **Naturalizing Cultural Values**

PART VI **Myth-Conceptions**

PART VII **Values and Biology Education**

LIST OF FIGURES

Prologue: The Ironies of Misplaced Assumptions

What seems more obvious than male and female as natural categories? Or that because genes govern our heritable traits, all the way down to cellular processes, they thus define our identity? Or that Darwin's principle of "survival of the fittest" aptly describes our fundamental selfishness in a competitive society? Or that biology typically advances by leaps of creative genius, epitomized by the rare achievements of our scientific heroes?

These notions are deeply rooted in our cultural psyche. Sometimes, they achieve an almost reverential status. We might call them "sacred cows." But because they concern biology and the nature of science, one might also entertain slightly more scientific terminology. I have playfully dubbed them *sacred bovines*. In the essays that follow, I extend that lighthearted spirit and invite you to reconsider these and other apparent truisms in biology. I hope you find the excursion enriching, as well as entertaining.

Our misplaced assumptions about biology are many. We may benefit by exposing them and learning how they shape our thinking and occasionally lead us astray. In addition, on a deeper level, we can examine all the errors as an ensemble. How did we come to misdirect our trust? How do our lapses reflect how we think, both individually and collectively? Musing on our errors offers the prospect of developing ways to think more effectively.

One great irony of evolution is how our brains, so well adapted in many respects, can indeed make mistakes. What we know—or what we think we know—is not always reliable. Our minds can err. It is a very human quality, expressed in the familiar saying "To err is human."

We are particularly prone to jump to conclusions. We tend to grab the first idea we encounter and run with it. So we fail to consider other, possibly better alternatives. We rarely wonder whether the pattern we perceive is incomplete or might reflect an unrepresentative sample. We rely just on the information that is readily at hand. Psychologists call it *availability bias*.[1] With more information, we frequently find that we were wrong.

Worse, from some perspectives, we also tend to insulate ourselves from recognizing, and thus fixing, these very mistakes. We use early impressions to filter later perceptions. We generate search images and models. Things that fit the pre-established templates are more readily noticed. We embrace evidence that confirms an original idea. At the same time, we mentally peripheralize cases that seem unfamiliar or strange. They rank as distractions. Similarly, we discount evidence that challenges an initial commitment. And all this generally happens subconsciously—without our "thinking" clearly about it. Psychologists refer to it as *confirmation bias*.[2] And most of them consider it the most prevalent source of error in thinking.

But the problem does not end there. Once we establish a particular notion, other ideas build on it, even if that notion is faulty. The original idea can become inextricably bound in a network of downstream reasoning. Call it *entrenchment*.[3] It does not occur just in individuals. Erroneous ideas can circulate socially—including in science. They become the basis for other ideas. In some cases, mistaken assumptions can become so widely used and so familiar that they seem virtually invisible. They take on the status of unquestioned truisms. In biology, they become sacred bovines. With perspectives so awry, any effort to correct the root error may itself be viewed as an error: a misguided assault on the plain, "established" facts. Remedying the error seems that much more difficult.

In short, typical patterns of human thinking lead to a cascade of problems. First, a tendency to think based on information that is encountered first or is readily available can eclipse more-thorough thinking. Confirmation bias can then reinforce mistaken ideas. Finally, entrenchment insulates them still further against being detected, identified, and corrected. Thus, our minds typically develop some significant blind spots. One might well wonder whether human brains are so well adapted after all.

Ironically, these tendencies to truncate thinking may themselves be products of natural selection.[4] Often enough, we just need *a* solution to a problem, not necessarily *the best* solution. Why waste needless time, energy, and resources? An organism that thinks "too much" may be less efficient; meanwhile, the early bird gets the worm instead. The way our minds take cognitive shortcuts—even though they may ultimately foster error—may be adaptive at an intermediate level.

Fortunately, perhaps, our animal ancestors also evolved the ability to learn. We may potentially discover an error, remedy the situation, and readjust our knowledge. Moreover, the experience may inform us on how to avoid committing that type of error again. We can learn from our mistakes. Our knowledge deepens. We improve our ability to negotiate an effective path through our ever-changing environment.

But how? If our brains are resilient and resist learning, how do we transcend our blind spots? How can we remedy the misperceptions permeating our sacred bovines?

Learning—or developing any truly novel way of thinking—often begins with a discrepancy. Some event disrupts expectations or interrupts normal functioning. An anomaly, or disturbing exception, in the fabric of understanding. A striking mismatch. Something just doesn't add up right. It's like a disharmonious chord of music in the mind: a cognitive dissonance.[5] The unanticipated event or unexplainable phenomenon arouses us from our mental complacence. We may well feel ill at ease. The case may engage our sense of wonder or spark our curiosity. The exceptions help dislodge old ideas and motivate us to find new ones. Puzzling contradictions inspire us to action: a critical role for *ironies*.

This book contributes to that spirit of discovery. It reveals and revels in hidden ironies. And in remedying errors. If you know where the blind spots are, you might accommodate them or engineer around them. The essays here invite you to reflect on some commonplace ideas that with further awareness may not seem so well founded after all. We can deepen our knowledge and appreciation of the world around us. We can learn to think more keenly.

This book contains twenty-eight short essays. Each profiles a separate sacred bovine. They fit into seven themes.

The first five essays address our commonplace notions about the nature of scientific practices, or how science works. They range from the perceived peripheral roles of emotions and political perspectives to the canonical formulation of the "scientific method" as a deliberate, stepwise process guaranteed to produce solutions.

A second set of essays (numbers 6–9) concerns unschooled impressions about the most central concept in biology: evolution by natural selection. They explore how Charles Darwin's biological concept of "survival of the fittest" relates to human culture. It is often construed unreflectively as a universal principle. For many, the concept implies that selfishness and competition in society are "natural" or inevitable, eclipsing any possibility of genuine ethics. The essays consider Darwin's own life, his ideas on the evolutionary basis of morality (less widely known than his hallmark theory), and the history of how others have interpreted his ideas. They also discuss current scientific views on the evolution of morality and the role of linguistic metaphors in perpetuating misconceptions.

The third theme (essays 10–12) concerns error in science. Science is ordinarily associated with a method guaranteed to generate the "right" answers. So it may seem hard to imagine that scientists make mistakes—or at least, any "justified" mistakes. Accordingly, the essays here explore errors by renowned scientists whose established achievements one could hardly discount. This include several Nobel Prize winners, the great Charles Darwin, and eighteenth-century chemist Joseph Priestley, who studied the role of plants in "restoring" the air "fouled" by respiration.

Next, three essays (numbers 13–15) consider occasions where claims regarded as scientifically valid by many members of the public are not endorsed by scientists themselves. Who speaks for science? Who is a scientific expert? Whose testimony on behalf of science can one trust? What counts as science outside the community of science proper? The question of negotiating whom or what to believe matters substantially in cases where individual or public welfare conflicts with economic or political interests, and where unscrupulous scientific "con artists" may appear.

In the fifth ensemble of essays (numbers 16–20) I examine several familiar cases of "natural" conditions and patterns. What may seem at first to be independent "nature" sometimes tends, on further inspection, to reflect or even embody human ideals. Cultural perspectives have been inscribed into our "objective" descriptions of nature, where they acquire a certain air of inevitability and purpose, often with the apparent warrant of science. The norms have been *naturalized*. These cases are among the most "sacred" of sacred bovines. Yet we may need to acknowledge that sometimes what one "sees" in nature reveals more about the seer than about what is actually seen. Scrutiny in these essays targets "human nature," male and female as unambiguous and exclusive categories, the concept of "normality" in organisms, genes as identity, and simplicity in nature.

The sixth set of essays (numbers 21–24) considers several widely celebrated scientific heroes. With greater context and historical detail, the heroic stories often seem far less heroic, even as the scientists become more vividly human. Why do we tend to monumentalize and idealize such figures, at the cost of damaging an understanding of how science really works? And why do such misleading images often function as role models? Are there alternatives to the conventional "myth-conceptions"?

The last theme (essays 25–28) addresses the intersection of biological knowledge with ethics and values. Science is typically regarded as independent of or free from the biasing effect of values. Yet facts also critically inform perspectives on values. In the instances discussed here, common cultural values can be better informed simply with more or better biological knowledge or with greater awareness of historical perspectives on science. Again, we might reflect on our customary selective awareness of science and history.

In a final epilogue, I review the journey through the many sacred bovines with an eye for general lessons about how to recognize misplaced assumptions. How can we nudge them back into view, where we can address them honestly? Yes, ultimately we can indeed foster habits for more creative and critical thinking. A separate afterword for teachers echoes these perspectives in a specialized educational context.

Challenging assumptions—our sacred bovines—can open opportunities. It spurs creativity and fosters insight—and deeper knowledge. Serious business, perhaps. But the process can also be fun. There can be sheer joy in enriching the mind. That sense of sly sprightliness is notably exhibited in the work of artist Andy Warhol. His vivid graphics of soup cans and pop idols like Marilyn Monroe or Elvis Presley helped awaken us to images that surround us in plain sight. We could see them differently. We could find the extraordinary in the ordinary. Likewise, American painter Jasper Johns, in his paintings of the American flag and other cultural symbols, alerted us to the emotive and cognitive power of such familiar icons. In a similar way, I hope to nudge perceptions and inspire reflections about biology. By highlighting the ironies in common beliefs, I hope to provoke learning about science. And a deeper appreciation of how our minds work. Maybe the plain and simple is not so plain and simple.

Ironically, left to themselves, our minds may blind us to the very truths we often seek. Our assumptions can be misplaced and so mislead us. But we can also exercise our minds to expose those hidden assumptions and help remedy them. That is the journey ahead. I hope you enjoy and find reward in exploring and reflecting on these sacred bovines.

PART I

The Way of Science

FIGURE 1.1 *Petrus Gonsalus, whose courtly dress reflects how he was valued as a "monster."*

1

Monsters and Marvels

Four-leaf clovers are traditional emblems of good luck. Two-headed sheep, five-legged frogs, or persons with six-fingered hands, by contrast, are more likely to be considered repugnant monsters, or "freaks of nature." Such alienation was not always the case. In sixteenth-century Europe, such "monsters," like the four-leaf clover today, mostly elicited wonder and respect. People were fascinated with natural phenomena just beyond the edge of the familiar. Indeed, that *emotional* response—at that juncture in history—helped foster the emergence of modern science. Wonder fostered investigation and, with it, deeper understanding of nature. One might thus well question a widespread but generally unchallenged belief about biology—what one might call a *sacred bovine*: that emotions can only contaminate science with subjective values. Indeed, delving into how "monsters" once evoked wonder might open a deeper appreciation of how science works today.

Wonder

Consider the case of Petrus Gonsalus, born in 1556 (Figure 1.1).[1] As one might guess from his portrait, Gonsalus (also known as Gonzales or Gonsalvus) became renowned for his exceptional hairiness. He was a "monster": someone—like dwarves, giants, or conjoined twins—with a body form conspicuously outside the ordinary. But, as his courtly robe might equally indicate, Gonsalus was also *special*.

Gonsalus was born on Tenerife, a small island off the west coast of Africa. But he found a home in the court of King Henry II. Once there, he became educated. "Like a second mother France nourished me from boyhood to manhood," he recollected, "and taught me to give up my wild manners, and the liberal arts, and to speak Latin."[2] Gonsalus's journey from the periphery of civilization to a center of power occurred because he could evoke a sense of wonder. Eventually, he moved to other courts across Europe. Wonder was widely esteemed.

For us, Gonsalus may be emblematic of an era when wonder flourished. In earlier centuries monsters were typically viewed as divine portents, or *prodigies*. Not that they were miracles. The course of nature seemed wide enough to include them. Still, why had the customs of nature been suspended in that way at a particular time and place? What purpose or intent did a monster signify? Why would this child, here, now, have such an inflated (hydrocephalic) head? Monsters thus once evoked fear or awe. The emotion reflected their uncertain meaning more than their strangeness of form.

By the 1500s, however, nature (still viewed as God's realm) seemed less capricious. Confidence in nature's consistency developed, although nature did not yet seem quite lawlike. The supernatural certainly still seemed possible: a divine power could suspend the natural order at any time. Monsters like Gonsalus were surely exceptional. Yet they seemed products of natural causes. That belief opened a new zone between the known and the unknowable. Historians Lorraine Daston and Katherine Park have dubbed such phenomena the *preternatural*, or "beyond the natural." The preternatural world, "suspended between the mundane and the miraculous," was emotionally charged. It was a domain of wonder and marvel.[3]

What did Europeans in the 1500s and 1600s marvel at? Magnetic attraction: How did it reach across empty space? The reputed power of the amethyst to repel hail and locusts. Invisible writing that magically reappeared when heated. Liquid phosphor in the sea near Cadiz. Gems emitting light. "Fool's paradises" of glass creating many colors from sunlight. Colored lights flickering in the northern sky. Healing a wound by bandaging the weapon that made it (if one believed that could work). Changing metals from one to another. An armor-plated cow-like beast with a huge horn on its nose. A sea-boar with tusks. A brainless child born in Montpelier. A child with a tail of a mammal. A woman with four breasts. Here was wonder indeed.[4] Monsters, in particular, reflected the intriguing tension at the edge of the natural: so close to human form, yet not. That is why Gonsalus—otherwise a wild "native"—found a home amidst the pinnacle of society.

Collecting and Exchanging

Objects that evoked wonder were worth saving and keeping. Such specimens were called *curiosities*. They included ostrich eggs, nautilus shells, whale vertebrae, a griffin claw (well, an inverted animal horn), armadillo shells, prickly blowfish, tropical corals, and dragons' teeth (probably sharks' teeth), as well as carved ivory and fossils. Add, too, exotic minerals and gemstones, oddly shaped bones, large turtle carapaces, and stuffed crocodiles. Curiosities, the physical artifacts, reflected the significance of curiosity, the emotion (as we would call it now).

The well-to-do, at least, began collecting curiosities. And they enshrined their treasures in special *curiosity cabinets*.[5] These cabinets allowed them to exhibit, take pride in, and perhaps share their unique specimens. In some cases, the collections expanded to fill whole specially dedicated rooms. All because of wonder.

Gonsalus fit into this cultural practice of collecting unique specimens. He was, perhaps, a living curiosity. In a sense, he was "collected" from his native Tenerife. He and other "monsters" who became members of court culture were unique "specimens," whose role was to elicit wonder. Gonsalus's uniqueness was ultimately documented and preserved in a full-length portrait. After 1583 it became a prominent fixture in the multiroomed curiosity "cabinet" of Archduke Ferdinand II of Tyrol (in Austria). Ferdinand's castle, Ambras, has since given its name to Gonsalus's condition: hypertrichosis universalis congenita, Ambras type.[6]

Ironically, perhaps, Gonsalus never owned his own portrait. While his uniqueness was valued, he was also essentially the king's *property*. We know that in 1595 Gonsalus's son Arrigo, who shared his striking hairiness (Figure 1.2), was given *as a gift* from Ranuccio Farnese of Parma, Italy, to his brother, Cardinal Edoardo Farnese.[7] While we do not share the ethical perspectives of the late Renaissance, we can clearly see how deeply the culture valued the sense of wonder.

Curiosities also contributed to Western European politics. Extraordinary specimens were exchanged as gifts among the rich and powerful, from one court to another. No mere gestures, these gifts were currency in establishing political alliances and seeking courtly favors. The more striking or rarer the specimen, the more valuable. As Arrigo's fate indicates, exchanges included living specimens. For example, an Indian rhinoceros, made famous by a 1515 Albrecht Dürer drawing, had been a gift from a sultan in India to the Portuguese governor there, who then gave it to the king in Portugal. It was on its way next to Pope Leo X in Italy when the ship carrying it sank. An elephant named Hanno had made a similar intercontinental journey, more successfully, the year before.[8]

FIGURE 1.2 Hairy Arrigo, Fool Pietro and Dwarf Amon, *by Agostino Carracci (ca. 1598). More monsters, more wonder.*

The demand for new marvels among the elite fueled a healthy trade. Merchants did not miss the opportunity to profit from venturing around the world. Curiosities, then, also became good business. Wonder supported commerce. Indeed, the commerce in exotica, combined with a spirit of dominion, helped finance voyaging and discovery farther from Europe. More and more specimens arrived as Europeans extended their political and economic domain. Collections expanded.

Wonder easily extended to the influx of strange new plants, new animals, new minerals, and new cultural artifacts. The diversity was exhilarating. The greater the diversity, the deeper the fascination. Many collectors now aimed for impressive scope as well. Collections expanded again, from select curiosities to comprehensive assemblages of thousands of specimens. One may readily appreciate how such collections and exhibit spaces evolved into natural history museums. The extraordinary collection of plants assembled by John Tradescant and his son, first catalogued in 1656, became the first public museum (Oxford University's Ashmolean Museum) in 1683. The collection of Hans Sloane, developed in the late 1680s, formed the core of the British Museum. In 1715 Albertus Seba, from Amsterdam, sold his collection of curiosities—one of the most extensive in Europe—to Peter the Great, who then created Russia's first natural history museum.[9] Wonder, an emotion, fostered the creation of vast collections that ultimately served more systematic study.

The Spirit of Investigation

Wonder was also important in spurring inquiry. One can easily imagine an emotional response to curiosities that is purely aesthetic and passive. Wonder, however, was not idle appreciation or, with monsters, a debilitating horror or awe. Rather, wonder was provocative. Strange specimens evoked questions about the natural order: What caused these forms to vary from nature's customs?

Wonder was also not curiosity, at least not at that time. In the 1500 and 1600s, curiosity implied a desire, even obsession, for knowledge that was inappropriate or unattainable. Curiosity was considered vain, self-absorbing, and indulgent. (Accordingly, dramatist Thomas Shadwell satirized Robert Boyle and Britain's young Royal Society as overzealous fools in his 1672 comedy *The Virtuoso*.) Wonder, on the other hand, led to fruitful investigation and to deeper knowledge about how nature worked (and often, too, to interpreting God's intentions). That motivation was central to the emergence of modern science.

The spirit of investigation typically manifested itself first in an effort to collect all that was known (or ever known) about a particular topic. Thus French surgeon Ambroise Paré expressed his interest in monsters by collecting information about as many cases as he could document and by reporting them all.[10] He categorized the various forms, combining explanations of natural means and divine intent: too much seed here, too little there, images impressed upon the mother's mind, maternal injury, hereditary illness, God's wrath (or his glory), and so forth. Paré's approach was encyclopedic although, by modern standards, somewhat credulous.

Human-animal hybrids appeared alongside conjoined twins and hermaphrodites. Wonder—at first—does not bring discernment. But Paré's work exemplifies well how such studies began, with few prior benchmarks. Eventually, the vast catalogs and natural history collections introduced two major challenges, each illustrated by Paré's experience and later addressed historically. First, how does one bring order to, or organize, everything, including all the unusual cases? Second, how does one distinguish credible from incredible claims? Both endeavors—the search for patterns and the development of standards for evidence—were important in establishing modern science.

The spirit of wonder and investigation was also nicely exemplified in the work of Ulisse Aldrovandi, of Bologna, Italy. Aldrovandi gained renown as one of the finest naturalists of the late 1500s. Like others, he collected specimens from the New World and around the globe. His collection, however, was among the largest and most amazing. By 1595 he could write, "Today in my microcosm, you can see more than 18,000 different things, among which 7000 in fifteen volumes, dried and pasted, 3000 of which I had painted as if alive."[11] Further, he set about recreating it all in a "paper museum" of illustrated books. He planned volumes on birds, fish, insects, trees, and—quite notably as an equivalent category—monsters.

The monstrous forms challenged Aldrovandi's organization. For example, he had collected many deformed lemons. Each was unique. Were they fundamentally different, or were they all "just" lemons? Aldrovandi gave each a separate category.[12] For him, the differences clearly mattered. Monsters were not just unusual regular specimens. They had a special meaning. Monsters were not easily classified; hence their continuing power to amaze.

Aldrovandi was also concerned about credibility, perhaps a bit more than Paré. He repeatedly remarked that his illustrations were drawn from life, rather than copied or based on some uncertain testimony. His books—his paper museum— drew on his actual specimens. He felt well informed enough to declare that one unicorn horn was a fake. And a purported mythological hydra, too. Also, based on one specimen he obtained, he showed others how a stingray could be reshaped to imitate a dragon (Figure 1.3)—although he still maintained that genuine dragons existed in nature.[13]

What about monsters? Were the reports and images of them credible? Could one really believe that somebody could be hairy all over, or, as Paré described one girl, "as furry as a bear"?[14] Petrus Gonsalus and his hirsute children, at least, would seem living proof. Aldrovandi surely appreciated the nature of material evidence. He found an opportunity to examine Gonsalus and his son in 1584 (one of only two physicians known to have done so). At one level, Aldrovandi sought perhaps to document Gonsalus as another anomaly for his collection. Yet he also tried to understand his unique condition. Elsewhere, Aldrovandi studied deformities in chick development and attributed them to chemical and physical changes in the egg yolk. Monsters shed light on nature. Here, wonder not only fostered the cultural practice of collection but also helped transform it into a deeper investigation of nature—what we now call science.

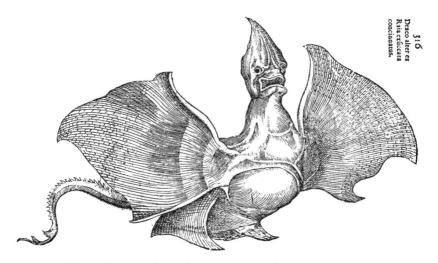

FIGURE 1.3 *Ulisse Aldrovandi's bogus dragon. Wondrous objects sometimes sparked questions about their credibility.*

Wonder Today

The haunting image of Petrus Gonsalus as both monster and courtier (Figure 1.1) reflects the central role of wonder in early science. A basic motivation for inquiry was, and perhaps still is, rooted in an *emotional* response. That lesson might encourage many young persons today to pursue a career in science. Science can be inspiring. Too often, it seems, school science—mandated by state curricula and broken into factoids on standardized tests—is lifeless, as dull as a textbook. But history affirms the pivotal role of emotions in fueling scientific practice. Science emerges from a sense of wonder and its sequel, an engagement in investigation.

The history also helps inform us of the critical boundaries of emotion. Too much astonishment, perhaps, and one risks generating an unproductive "gawking wonder," not inquiry.[15] Television "discovery" programs and science centers, in particular, can suffer from a superficial "ooh-aah" syndrome. The excitement they generate can be fleeting. The emotion quickly fades away. When they do not engage the viewer in deeper probing, science seems more a circus sideshow than a research endeavor. Authoritative answers and dumbstruck awe substitute for the adventure of pursuing questions. There is an art to motivating fruitful questioning.

One can easily find cases of "monsters" like Gonsalus in books and on the Internet.[16] But our modern responses to these cases often differ substantially from those in the sixteenth and seventeenth centuries. Monsters have mostly become freakish, evoking disgust, not wonder. (For more on that puzzle, see essay 17.) Wonder emerges by challenging commonplaces. For example, the hermaphrodites that fascinated Paré and his peers can still potently problematize our notions of male and female (essay 16). As a modern example, situs inversus, the lateral reversal of internal organs,[17] might also disturb our sense of the natural order. Nonhuman examples, such as albino gorillas or tigers, bald lemurs, and other anomalies occasionally found at zoos, may likewise prompt queries about genetics and development. Wonders engage us and keep us off balance.

In modern culture, science typically serves as an icon of objectivity, purged of emotion. One might thereby imagine that scientific practice, too, should be sterile, lest subjectivity intrude and poison the science. We may have to rethink that sacred bovine, however. Understanding the historical era that made hairy Gonsalus special indicates that the emotion of wonder may well be constitutive of science. Will wonders never cease? Let us hope not.

2

Ahead of the Curve

Graphs function plainly to summarize data. They hardly seem momentous. They are not like a famous discovery, whose significance is often marked by an eponymous name: *Mendel's* laws, the *Watson and Crick* model of DNA, *Darwinian* theory. Who would name a mere graph? They seem mundane fragments of science, hardly worth celebrating. A notable exception, however, is the Keeling Curve (Figure 2.1). This simple graph depicts the steady rise in the concentration of carbon dioxide (CO_2) in the Earth's atmosphere over the last half century. It helps document how humans have transformed the atmosphere and, with it, the Earth's temperature. The Keeling Curve is a linchpin in the evidence that humans have changed the planet's climate.

The Keeling Curve starts in 1958 and continues uninterrupted for over five decades. The scale of the data is extraordinary, an ideal rarely achieved in science. The hard data from real-time measurements show the steady accumulation of CO_2 from burning fossil fuels. The results, presented in a simple yet striking

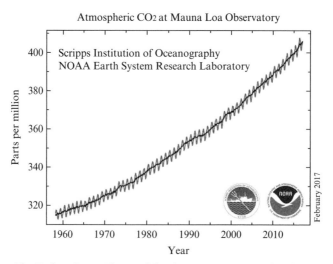

FIGURE 2.1 *The Keeling Curve. This graph has become an important benchmark in our understanding of global warming, but how did it originate?*

visual format, serve to warn an energy-hungry culture of its environmental hubris. Although just a graph, it is monumental in scope and significance.

The Keeling Curve, viewed in retrospect, raises an interesting question about how science works. How do such important long-term data sets emerge? Often we assume that scientific investigations find just what they intend to find. That is an implicit lesson of the tidy scientific method, as widely presented (see essay 5). But should we trust this sacred bovine? Could anyone have predicted this curve or its importance in advance? How did these important data *originate*? What happened *before* the graph was fully created? What happened, literally, ahead of the Curve?[1]

A Measured Approach?

The Keeling Curve is named after its creator, Charles David Keeling. In the 1950s, as a handsome young man frequently enjoying the great outdoors (Figure 2.2), he hardly fit the stereotypical image of a scientist clad in a white coat, isolated in a lab. Indeed, with a fresh degree in chemistry, he turned down many job opportunities because he wanted to be closer to nature on the West Coast. As an initial project

FIGURE 2.2 *David Keeling in a personal family photo from 1958, around the time he started collecting global data on atmospheric CO_2.*

in his new position, he focused on how to extract uranium from granites for use in nuclear power. After two weeks of crushing rocks, however, Keeling felt uninspired.

Then he overheard a small geochemical puzzle. Could one determine the carbonate level of surface water by assuming it was in equilibrium with the carbon dioxide in the air above and the carbonate rocks below? Soon, Keeling was driving up the dramatic Big Sur coast of California, camping amid the redwoods, and waking every few hours at night to collect air, river, and subsurface water samples. Back home, he rigged up an apparatus to measure the scant amounts of carbon dioxide in his air samples (less than 1%). He needed some technical expertise to secure precise and reliable measurements. Keeling's results indicated that the original idea was ill founded. The concentration of carbon dioxide was much more concentrated in the water than in the air. But he also noticed that the CO_2 concentration of the air was relatively constant, although rising at night. Keeling now shifted his focus to measuring just the atmospheric CO_2.

At the time, measurements of atmospheric CO_2 ranged widely. Scientists thus assumed that it varied from place to place, as a result of unknown local factors. Now intrigued, Keeling wondered if there was a consistent baseline and an identifiable pattern to the variation. Over the next year he collected samples from more than a dozen sites across the West Coast—while appreciating the mountain, desert, forest, and seashore scenery along the way. Keeling found that CO_2 levels reflected the density of local vegetation but otherwise seemed surprisingly constant, especially where winds mixed the air well. He was now convinced that there was a standard level of atmospheric CO_2 globally.

An opportunity to expand his studies emerged the following year. As the Cold War took hold, the United States wanted to develop more knowledge about the Earth—That might enable one, for example, to detect enemy submarines in the oceans or gauge the possible effects of nuclear weapons. The project was all framed in the guise of international cooperation: the International Geophysical Year, 1957–58. The military was also concerned about whether carbon dioxide in the air would absorb heat and divert heat-seeking missiles. So the Weather Bureau was already set to measure CO_2. Keeling's proposal to measure the levels of carbon dioxide was well received, and he was able to establish remote stations in Antarctica and at the windswept peak Mauna Loa, in the middle of the Pacific Ocean, as well as at the Scripps Institution in California. Roughly two years later, Keeling still had no consistent CO_2 measurement. Instead, he had documented a seasonal flux in CO_2, due to the deciduous forests in the Northern Hemisphere.

Keeling wanted to continue. He secured money from the still young National Science Foundation (NSF). The recent Soviet launching of the satellite Sputnik had fueled US anxieties about losing political superiority, and more funding for science was readily available. But two years later, the funds ran out. The Mauna Loa Observatory was forced to close. Keeling traveled to Washington, DC, and was able to persuade NSF to resume his funding. By 1964, Keeling's CO_2 data were drifting upward. But it was difficult to see the modest change as a solid trend. It might have reflected some natural cycle, say, in solar activity. Keeling had become aware of carbon dioxide as a greenhouse gas and its potential for warming the planet. He presented his data at a conference on the history of climate, but everyone seemed far more concerned about the possibility of an impending new ice age than of any increase in temperature in some remote future.

By the late 1960s, however, the trend of rising CO_2 was becoming unmistakable. After ten years of data, the annual minimum had exceeded the first maximum, from 1958. What no one could say definitively at the time was how the rise might affect global temperatures. First, the oceans seemed likely to buffer any changes. Yet with public concerns about the environment on the rise, a presidential advisory committee recommended that scientists continue to monitor carbon dioxide levels and study potential warming. The environmental significance of measuring CO_2 had become established.

But the recognition of its scientific importance hardly meant that funding was guaranteed. Over and over again for the next two decades, Keeling faced threatened shutdowns of his measurement program. First, a newly reorganized government agency started measuring CO_2 on its own, implying that it would replace Keeling's own work. But Keeling did not trust the quality of its measurement techniques. He continued his own work independently. Then the NSF cast his work as "routine" and vowed to withhold further funding. So Keeling scrambled to find new ways to use the data, revealing unexpected patterns related to tropical weather systems. The government, hoping to reduce costs, tried to institutionalize less stringent measurement methods. Whereupon Keeling rallied international support that firmly established his more rigorous standards. The government and NSF continued pressure to take control of the program. So Keeling found yet new ways to generate novel discoveries from the data. He showed how historical ratios of radioactive carbon specifically implicated fossil fuels in the CO_2 increase. Keeling continued to worry that switching to poor-quality government measurements would invalidate long-term analyses of the data. Then, under the new presidency of Ronald Reagan, spending for environmental science was cut unilaterally. Hearings by a junior congressman, Al Gore, however, helped raise public awareness of global warming, and some funds were restored.

Ultimately, Keeling managed to secure continuous funding from the Department of Energy. Over the course of three decades, however, Keeling had relied on at least nine sources of funding, most lasting only a few years. Given the challenges of funding and struggles over measurement standards, one might wonder how a long-term data set could have been built at all. Ultimately, it was. The mere existence of the measurements on this large scale is an implicit tribute to Keeling's perseverance and his political and scientific creativity. It is indeed fitting that we call the result the Keeling Curve.

Science, Backward and Forward

Thinking backward from today's Keeling Curve, one might easily imagine that Keeling had some special genius in knowing how his measurements would be meaningful decades later. If one views science as a methodical unfolding of inventing and testing theories, that may seem the only explanation. Keeling seems a visionary, "ahead of the curve," as the saying goes.

One might instead take an interpretive clue from evolutionary biology. Evolution, we know, is nonteleological. That is, short-term changes are not guided by some ultimate purpose. Natural selection acts on adaptive features in the moment, not an imagined future. Lightweight feathers are now integral to flight. But they originated for a different reason. Long before the prospect of flying, they insulated dinosaurs. Likewise, lungs made the vertebrate transition to land possible. But originally they helped regulate buoyancy in organisms that were thoroughly aquatic. Swim bladders became lungs when the environment changed. Similarly, ears enable all sorts of behaviors associated with sounds in the air, from escaping predators to finding mates. But ears are evolutionary remnants of the lateral line organ in fish, which detects nearby movement in the water. Contexts change. Functions transform. Current function can mask the history. Evolutionary changes indicate the need to conceive history from the perspective of the past, proceeding forward.

So, too, with the history of science. One would be mistaken to think that because Columbus "discovered" America for Europeans, he had foreseen or intended the momentous migration of European population and culture that followed. Science inches forward, somewhat blindly perhaps. It depends on chance and unexpected contingency. Ultimately, it is not possible to be "ahead of one's time." Keeling's recognition is well earned, but not for an early insight into an unpredictable future. Following the history of the Keeling Curve looking forward shows how contexts can shift and observations can be recontextualized.

The very phrase "ahead of the curve" has its own, ironic history.[2] Nowadays, it has come to share a meaning with "ahead of the game" or "ahead of the pack," in the sense that someone seems (in retrospect) to have exhibited leadership by anticipating future events. For some people, the "curve" is the bell curve. In that context, "ahead of the curve" means rare, exceptional performance, like "head of the class." But the phrase originated in aviation. Airplanes deal with both lift and drag, each based on airspeed. The relationship between airspeed and drag, marking the transition between falling and flying, is known as the power curve. To maintain control of the aircraft, a pilot wants to remain "ahead of the power curve." Sometime in the 1970s, apparently, the phrase jumped through military analogy into political contexts where White House administrators wanted to maintain control as the public reacted to adverse news. Now, the history of the phrase is, well, history. And we blithely forge ahead, imagining on occasion that someone might be "ahead of the curve." The history of discovery looks different, backward and forward.

3

Marxism and Cell Biology

> There is no royal road to science, and only those who do not dread the
> fatiguing climb of its steep paths have a chance of gaining its luminous
> summits.
>
> —Karl Marx, *Das Kapital*

Few biologists today have likely heard of cell biologist Alex Novikoff (1913–
1987) (Figure 3.1). But the fruits of his science are well known. He helped
discover the cell organelle called the lysosome. In 1955 he visualized what
Christian de Duve had characterized only by chemical means. He documented
the first known enzyme of the Golgi body, another cell organelle. He developed
ways to stain lysosomes and peroxisomes (also cell organelles) that were critical
to identifying them and studying them with the electron microscope. Novikoff
also was targeted by the anti-Communist movement in the mid-twentieth century. In 1953 he was dismissed from the University of Vermont for declining
to answer questions before a congressional committee. In 1974 he was elected
to the National Academy of Sciences. His FBI file then contained 822 pages.
Novikoff's fascinating case raises important issues about how science and political ideology relate.[1]

In 1982 the American Society for Cell Biology honored Novikoff with its prestigious E. B. Wilson Award for his foundational contributions to the emerging field.
Yet much earlier, in the late 1930s, he was indeed a member of the Communist
Party. For him, it expressed a quest for social justice and an appreciation of Karl
Marx's scientific posture toward society. While he researched experimental embryology as a PhD student at Columbia University, he also helped write and distribute
the Communist newsletter at Brooklyn College, where he taught. When the college tried to disrupt the teachers' union, Novikoff was secretly listed as a suspected
Communist. When World War II began, Novikoff wanted to serve the nation. He
applied for a medical commission in the military. He was twice denied, however,
owing to doubts about his loyalty. He later consulted for the army on two biological films—until it found his vague Communist record. (One wonders: Did someone
imagine that he could link enzymes and carbohydrate metabolism to the violent
overthrow of the US government?) Later, Novikoff lost his faculty position—not
for any political activity but for invoking the Fifth Amendment in anti-Communist

FIGURE 3.1 *Alex Novikoff at the electron microscope of Albert Einstein College of*
Medicine in 1955. Marxist ideas shaped some of his important discoveries in cell biology.

hearings, and despite recommendations from fellow faculty describing his "tireless"
research efforts. In short, cultural politics adversely affected Novikoff's scientific
career.

Novikoff's case may seem at first to epitomize how science and politics don't
mix. Yet it also illustrates vividly that science is not wholly insulated from culture,
despite popular images of science as pure and transcendental. How are the two to
coexist? The case may seem to confirm a widespread impression that politics can
only interfere with the proper conduct of science. Still, one may profitably consider
the assumption—another sacred bovine?—that science is best sheltered from political ideology.

Novikoff's story would not be nearly so engaging—nor perhaps merit much further comment—if his politics did not also positively influence his work. Traces of
Marxist ideas permeate his conceptual outlook and his interpretations of evidence,
as well as his professional conduct. Cell biology, of all subjects, may seem incred-
ibly remote from politics. Yet careful consideration of Novikoff's work shows how
political ideology may sometimes contribute fruitfully to the practice, and even the
content, of science. Ultimately, such analysis might deepen reflection on the nature
of science and what ultimately makes knowledge reliable.

Dialectics and Cells

Novikoff came from a poor, struggling immigrant family. An outstanding student, he graduated from high school at age fourteen. When he completed undergraduate studies at Columbia University, institutions had begun limiting opportunities for Jews. Despite stellar credentials, Alex was not admitted to medical school. Embittered, he developed an uncompromising advocacy for social justice. Those feelings were rekindled when he began teaching at Brooklyn College in 1931 and encountered disparities between junior and senior faculty, with no consideration for ability. For Novikoff, Communist ideology offered prospective solutions to what seemed an unjust power structure.

By 1945 Novikoff, age thirty-two, was married and had formally severed ties with the Communist Party. But Marxist patterns in his thinking remained. In one paper that year, Novikoff profiled the problems of extreme reductionism, on the one hand, and implicit vitalism, on the other. In the first view, life was explained merely by the mechanical combination of its parts. The alternative was to view life as exhibiting some special property that explained its unique organization. Novikoff, by contrast, underscored the need to understand parts and wholes together, using a "dialectical approach."[2] The concept of *dialectics* was from Marx. For Marx society exhibited a struggle between two economic groups: those who owned capital (and thus held power) and laborers. Their conflict would be resolved—and workers freed—only by creating a new system that completely dissolved the owner-laborer relationship and integrated everyone at a new level of communism. Philosophically, this action by the disenfranchised would represent a synthesis of opposing perspectives. That was Marx's essential dialectic: progress emerging from integrating opposites, as modeled in political history. Novikoff echoed that concept and language. He argued for reconciling the two opposing biological views in what he called "the concept of integrative levels in biology."

Novikoff privileged neither atomism nor organicism. Nonetheless, he reserved his strongest criticisms for the reductionists. He reminded readers that to understand cells fully, biochemistry, however essential, was not enough. One also needed to know cell structure. Likewise, isolated cell functions alone were not physiology. One needed the developmental context. Novikoff's primary concern, however, was interpreting society—and doing so scientifically, as Marx advocated. He decried the misleading organism-society analogy, as proposed by Herbert Spencer and his followers, that reduced culture to biology (see essay 7). Conflating levels, he claimed, could lead to "erroneous and dangerous social conclusions." As an example Novikoff pointedly identified the fascists (Nazis), who alleged that "man's biology decides his social behavior."[3] Genetics, Novikoff observed, was incomplete. There was also sociocultural inheritance. Sociological principles were needed, he maintained, to keep society "free and democratic." Here, his antireductionist science was clearly informed by political ideology.

Novikoff continued to apply Marx's dialectical approach fruitfully over the next three decades, even to his investigations of cells. In a reductionist vein, he localized biochemical functions to parts within cells. But he also did not lose sight of context or more-holistic perspectives. He studied lysosomes and peroxisomes in diverse cell types and tissues and in pathological conditions (such as fatty liver, tumors, and nephrosis). That work revealed how the "same" units differed in various cellular contexts, or wholes.

In 1965 James Watson published *Molecular Biology of the Gene*. Novikoff saw in it an unproductive molecular bias. In 1970 he aimed to remedy that bias "dialectically" in co-authoring his own text *Cells and Organelles*.[4] It was one of the first textbooks of cell biology, widely used through three editions. Here, parts and wholes received equal attention. After descriptions of the many organelles (as parts), Novikoff included just as much coverage on the many cell types made from combining the parts into different wholes. An unstated Marxist perspective high-lighted how cells (like societies) did not reduce to simple sums of their independent component parts.

Novikoff's political perspective led to an appreciation of how parts related in forming higher levels of biological organization. For example, while lysosomes, per-oxisomes, and endoplasmic reticulum were not considered "dominant" organelles, he recognized the implicit significance of their "labor" to the whole and devoted study to them. Novikoff was also sensitized to see differences in parts, with their import for integrated wholes. As reported in an important 1953 paper he separated broken cells into ten layers (or particle sizes), rather than the customary four. That allowed a finer-scale analysis. He tested each division for the activity of seven care-fully selected enzymes, then modified the separation process, matching particular particle sizes with the corresponding enzyme activity. Ultimately, he mapped the characteristic enzymes to six organelles, two not yet known. His innovative method led to discovery.

In a similar way, he examined liver tissue, demonstrating that cells assumed to be all the same were actually different, based on both biochemistry and cell struc-ture. Studying peroxisomes with an eye to differences allowed Novikoff to discover microperoxisomes in 1972. He also noticed a close association of the *G*olgi body, *e*ndoplasmic *r*eticulum, and *l*ysosomes, a hybrid structure he called GERL, which helped clarify how lysosomes originate. Novikoff's discoveries were guided by a conceptual map of what merited notice. And that outlook had been influenced by Marxist ideology.

Materialism and Evolution

The 1945 paper on integrative levels exhibited another core Marxist principle: mate-rialism. Novikoff noted that many nonreductionists appealed to various guiding forces in evolution, such as an "organizing trend," inherent progress, or directional (orthogenetic) trajectories. Today, such forces are well outside sound biology, but biologists in the early twentieth century had used them to explain the history of fos-sils. Novikoff criticized such nonmaterialistic forces as unsubstantiated and super-fluous. Here, he echoed Marx's view of history. Marx saw how concrete economic relations shaped society, as well as political changes, through history. He thus regarded intellectual movements as responses, not causes. He emphasized instead the material causal elements of history. Accordingly, Marx viewed Darwin's theory of descent with modification quite favorably. It described organic change in material terms. For Marx, natural history and human history were parallel. That materialist framework enabled Novikoff to identify the biological weaknesses of certain evolutionary concepts.

Novikoff's discussion had political overtones, as well. Many followers of Spencer claimed that progress was "natural" and that humans should not disturb it, lest it cease. They promoted a laissez-faire approach to society (essay 7). From a Communist perspective, Novikoff could easily see that their argument served only to preserve the status quo and thus protect those already in power. He regarded such trust in progress as unsound fatalism. Rather, "social progress," he declared, "rests upon the planned activities of men."[5] In a sense, he set political action in a biological context.

Communist ideology aimed to inspire laborers to act to change history materially. Novikoff echoed those sentiments in portraying the evolution of humans and their traits. He observed that humans are able to control their environment: they are not bound to their political history. Their intelligence, likewise, is plastic: political change is possible. He further commented that "man possesses a unique head and hand," alluding in particular to the power of manual laborers.[6] Marxist ideology gave special relevance to certain features of evolution that remained obscured under other perspectives.

Novikoff also viewed science as a form of work. In the opening and closing chapters of his textbook, he profiled the history and methods of cell biology. Science was not static information but an active, engaging field that the reader might consider pursuing. Similarly, in a children's book on physiology, Novikoff described many historical scientists, presenting science as a human endeavor. Novikoff himself was extremely active. He worked with little sleep. He wrote most of his PhD thesis on a train while commuting to and from school and work. In 1955, he collaborated with de Duve on lysosomes. Because no electron microscope was available in Louvain, Belgium, at the end of each day he packed up the cell samples in an iced thermos bottle, took two trains to a lab in Paris, and continued working into the night to produce images, returning to repeat the routine again the next day. Colleagues repeatedly described Novikoff's efforts as "tireless." He certainly exemplified the principle valuing material work.

Socialism

Finally, one may note how Novikoff's Marxist orientation affected the practice, not just the content, of his science. His socialist ideals were expressed in many ways. First, Novikoff was one of the first to write science books for children. *Climbing Our Family Tree*, an introduction to evolution, is a landmark in children's literature.[7] Novikoff, not yet a parent himself, clearly saw even young readers as important.

FIGURE 3.2 *Illustration depicting the Marxist ideal of human freedom and potential for progress in an evolutionary context, from Novikoff's book for children,* Climbing Our Family Tree.

So too, presumably, did the Communist press that published the book. The text is informative while also highlighting Marxist themes (see Figure 3.2). Novikoff dramatized evolutionary innovations—such as homeostatic internal environments, the transition to land, internal development, and homeothermy—as steps in organisms' becoming freer from their environment. The final chapters introduced human society, too, as a product of evolution, and underscored the basis for culture and cooperation. "Men, working with each other," he concluded, "can become ever more free—ever more human."[8]

Second, Novikoff viewed the scientific community as one of equal peers. (Not all scientists do.) He generously shared credit for work done. Nearly all his published papers are co-authored. More deeply, he was open to conversing with anyone, much to the surprise of students and junior scientists. Such discourse promotes the exchange of ideas. Given that Novikoff endorsed dialectics, he also enjoyed the opportunity to vigorously debate alternative views with anyone—yet graciously conceded when shown to be wrong. That "socialist" spirit contributed to his critical analysis and thereby to more-robust scientific conclusions.

Interpreting Politics in Science

Not all influences of politics on science are positive, surely. For example, one may readily point to Lysenko's notorious suppression of Mendelism in the former Soviet Union. Given Novikoff's case, however, one must qualify any universal negative claim. One may be tempted to consider Novikoff as "just a good biologist," his achievements wholly unrelated to his politics. However, such a conclusion blindly disregards how Novikoff came by his skills. The language of his 1945 publications, in particular, leaves no doubt about his conceptual roots. To explain Novikoff's scientific achievements fully, his personal history and Marxist perspective are essential.

Novikoff's science was science in part because it was not exclusively Marxist. Nor did Novikoff ever present political arguments to justify his conclusions (as Lysenko did). Rather, his political ideology functioned in what philosophers of science sometimes call the "context of discovery." Accordingly, Novikoff's ideology was a valuable tool for generating alternative approaches or probing possible interpretations. Novikoff could sometimes appreciate what others, with their own conceptual blind spots, could not. Beyond discovery, science also relies on a complementary "context of justification," where standards of evidence apply. Novikoff exercised his standards here, too. Overall, science evolves through an integrative coupling of blind variation (discovery) and selective retention (justification), akin to the familiar biological process of adaptation through natural selection.

Novikoff did indeed respect arguments for alternatives. In his 1953 study of cellular enzymes discussed above, for example, he first attributed the observed differences to known organelles, the mitochondria and microsomes (ribosomes). De Duve learned of the results, and the two later met in New York and chatted in Central Park. De Duve shared his results indicating the presence of an undocumented organelle, the lysosome. Novikoff accepted de Duve's (re)interpretation of his own results. The two went on to collaborate briefly and became lifelong friends. Political ideology can be productive, so long as one still listens to criticism and minds the evidence.

Ultimately, Novikoff's Marxist political perspective enriched science. Ironically, such real—and fruitful—influences were never the concern of the anti-Communist demagogues who dogged Novikoff for the allegedly subversive consequences of his political views.

4

The Messy Story behind the Most Beautiful Experiment in Biology

"The most beautiful experiment in biology." That was how John Cairns described it: the 1958 experiment that showed how the genetic material, DNA, replicates. The work is still widely celebrated, sometimes in introductory biology textbooks.[1]

This esteemed experiment by Matt Meselson and Frank Stahl (described more fully below) and others like it reflect an ideal in science, one marked by an intuitive aesthetic response. The test was simple. The results were clear. The method and reasoning seemed obvious. Theory and evidence complemented each other elegantly. That seems to be how science works—or should work.

However, this view of biology, so common as to be beyond question—another sacred bovine?—can be misleading. Appearances can be deceptive. Delving into the history of this now-famous experiment fosters a very different image. Behind the apparent simplicity hides extraordinary—and fascinating—complexity. A glimpse of the messy world of investigation indicates how science really happens, quite apart from the tidy scientific method that one finds in standard textbooks. Ultimately, the messy story behind the most beautiful experiment in biology offers a quite different, and deeply informative, way to appreciate science.[2]

A Beautiful Experiment

The experiment developed from a puzzle about how DNA, the genetic molecule, replicates. In 1953 James Watson and Francis Crick, building on data from Rosalind Franklin and Maurice Wilkins, presented a model of DNA's molecular structure. It was two threads that coiled around each other, they claimed. Like two intertwined strands of rope. That double helix model has since been widely celebrated and inspired much art.

But how did the DNA molecule replicate? When any cell divides, each new cell receives a complete set of information. Duplicate copies of DNA are assembled. Watson and Crick had only hinted at how that might occur. The genetic information was a sequence of units, called nucleotides, that bridged the two strands. They occurred in pairs. The shapes in each pair were complementary. So the shape of one side would determine which missing base would pair on the other. Units could self-assemble blindly. But one detail remained unclear. Did the original molecule act as a template for making two new strands that joined with each other? Or did the original split, with each half pairing with its own new strand?

While the alternatives are easy enough to envision with cardboard cutouts or handheld models, the challenge was to make such events on a molecular scale visible experimentally. That is just what Meselson and Stahl did. They solved the puzzle, showing in 1958 that DNA replicates "semi-conservatively," distributing the original DNA strands half and half in each cell generation.

The solution involved two experimental innovations. First, Meselson and Stahl invented a way to identify the new versus parent strands. Second, they developed a method to separate the different forms of DNA resulting from successive replications. The new strands were assembled using atomic isotopes with a modestly heavier molecular weight. Once suspended in fluid, heavy and light strands could then be separated by spinning them at high velocities in a centrifuge. They used a heavy salt solution that, as the tubes spun rapidly, generated a density gradient. Heavy molecules would sink, and light molecules, by contrast, would "float." Each molecule (based on its weight) would drift and at equilibrium find a distinct level in the gradient. The resulting bands of DNA material at each generation were visually definitive, or as celebrated by one researcher, "perfect Watson-Crickery."[3]

The experiment epitomized good practice in several ways. First, it captured a central theoretical question in a single experiment. The problem of DNA replication was certainly not new. Watson and Crick's model had puzzled researchers for several years. Imagining possible events at the molecular level is relatively easy. Manifesting them in the lab is quite another thing. Sometimes, the molecular realm is revealed piecemeal, in clues and partial glimpses. Here, one well-oriented probe sufficed.

Second, Meselson and Stahl's experimental design addressed all the alternative theoretical models of replication simultaneously. If they failed to confirm one model, they would not have to then test the others.

Third, the experiment was marked by laboratory expertise. Material skills matter as much as thinking. The results were clean and unambiguous.

Finally, the team introduced a new method of wide scope. The technique of using heavy isotopes to differentiate macromolecules, once demonstrated, could be applied to many other studies.

Meselson and Stahl's experiment thus exhibited creative arrangement of laboratory conditions, theoretical import, clarity, and craft skills, all while pioneering an important new method. Rarely do all such elements come together in one work. When they do, scientists justly celebrate.

Typically, we associate beauty with works of art or design. Yet our aesthetic sense of unity and harmony responds whenever form and function match. Scientists thus come to regard some experimental designs as elegant. They appreciate how a set of observations may be specially configured to yield a decisive interpretation. No procedure is unnecessary, no effort wasted. Stahl himself later commented on the perceived beauty in his experiment: "It's very rare in biology that anything comes out like that. It's all so self-contained. All so internally self-supporting. Usually, if you are lucky to get a result in biology, you then spend the next year doing all those plausible controls to rule out other explanations; but this one was just a self-contained statement."[4] Meselson and Stahl's experiment was beautiful because the experimental methods and theoretical results fit together so fully, yet so minimally. No wonder that the work is sometimes featured in biology textbooks, where it can exhibit scientific ideals.

A Messy History

The ultimate outcome was beautiful. But how did it emerge? How do such notable achievements in science unfold? How did Meselson and Stahl originally conceive their novel experiment? What shaped its groundbreaking conditions? Indeed, how do scientific discoveries happen? Historian Larry Holmes documented the episode by examining lab records, interviewing the scientists, and plumbing institutional details. Viewed as a creative process, the experiment proved to be far from simple. Ultimately, the account filled a 500-page book.[5]

Matt and Frank met as graduate students in the summer of 1954 while at Woods Hole Biological Laboratory (Figure 4.1). Matt was a course assistant for James Watson himself. Frank was taking another course, not available at his home institution. One afternoon Frank was drinking a gin and tonic under a tree. Watching from the main building, Watson remarked on his reputed fine lab skills. Matt, curious, went to introduce himself. Frank had been considering a statistical problem requiring calculus. Several days later, Matt offered a solution, impressing Frank in turn. An enduring friendship developed. Before the summer was out, Matt had

FIGURE 4.1 *Matt Meselson (left) and Frank Stahl (right). The wandering history of their friendship led to what has been called "the most beautiful experiment in biology."*

mentioned a prospective study on DNA replication, and Frank had structured it experimentally.

Where had Meselson's idea come from? Early in 1954 he had been working on problems on deuterium, a heavy isotope of hydrogen, as part of a course with Nobel Prize–winning chemist Linus Pauling. He wondered whether organisms would live in "heavy water," made with the heavy hydrogen isotope. Later that spring, a visiting lecturer, Jacques Monod, had spoken on how cells can be induced to produce new enzymes. Meselson then imagined how he could grow cells in a medium with the heavy isotope, which would then incorporate itself into the new proteins. One could then separate the new proteins from the old in a solution with an appropriate density. The old proteins would float to the top, while the new, slightly heavier ones sank to the bottom. The core design for the later experiment on DNA was thus first developed in an entirely different context.

A few months later, a molecular biologist, Max Delbrück, introduced Meselson to his ideas on DNA replication. He speculated that the original paired strands of DNA would remain intact and that replication would occur in short isolated segments. Old and new fragments would thus likely be interspersed, spliced into patchwork strands.[6] Meselson saw another application of his heavy-isotope scheme—the one he would share with Stahl at Woods Hole in the summer of 1954.

But the route to the last run of the experiment in February 1958—over three years later—was hardly direct. Before long, the heavy deuterium was replaced by 5-bromouracil, very similar to the thymine unit in DNA. It would substitute directly for thymine during DNA synthesis. The weight difference of old and new strands would be greater, and thus separating them would be easier.

This idea in turn led the team to a wholly different line of investigation, on using 5-bromouracil to induce mutations, which could then be studied in more detail. Soon the second question became primary. Pursuing that new trajectory, they searched for a solution with an appropriate density. Would potassium bromide work? No. Magnesium sulfate? No. Barium perchlorate? No. Cesium chloride? Perhaps. But at what concentration? Each additional variable required more brute trial and error. Then they tried the new technique, only to discover that centrifuging the solution destroyed the homogenous density. To their dismay, they produced a density gradient instead. They had planned to separate the DNA in *discrete* layers using successive solutions of uniform known density. Fortunately, the gradient was gradual enough. The key molecules would still separate and spread themselves out according to their weight. Nonetheless, they explored a possible alternative: separating the molecules in a gel medium subjected to an electric field. One method led to another, and another still: lots of trial and error.

In August 1957, Meselson saw an advertisement for an isotope of nitrogen, nitrogen-15. They had rejected it earlier, assuming that the weight differential of DNA using nitrogen-14 would be too small to measure. However, the unexpected resolving power of the density gradient method now made it possible to return to the original nitrogen isotope. Suddenly, 5-bromouracil was abandoned. The mutagenesis inquiry was set aside. The experimental design so celebrated by history finally emerged.

But there were other practical challenges as well. The two researchers competed for time on the centrifuge, a huge (and costly) instrument shared by many labs. A sample run might take twenty hours or more. Then they had to find the appropriate spinning speed, finally settling on nearly 45,000 revolutions per minute, producing a force over one hundred thousand times the force of gravity. They needed to identify a chemical to break open the bacterial cells without also destroying the DNA molecules inside. Meanwhile, their regular work continued. Meselson had a doctoral thesis to write—on an unrelated topic. There were job interviews. And so on. If the final experiment was simple, the process was anything but.

Real science hardly resembles the cookbook labs one frequently encounters in school. Nor is it the formulaic scientific method enforced on many science fair projects. Of course Meselson and Stahl's investigation may be edited and reconfigured in retrospect to fit a simple step-by-step logic from hypothesis to experiment, results, and conclusion. But the historical view, in real time, reveals a far less orderly or directed history. The outcome was beautiful. However, the process was messy.

A Hidden Reward

Science is filled with chance encounters, metaphoric thinking, false starts, tinkering, unsuccessful explorations, and plain hard work. As illustrated by the convoluted history of Meselson and Stahl's experiment, science is a creative enterprise that collects isolated moments of imagination and available resources and knits them into multiple possible trajectories. Ultimately, unlikely arrangements of methods and observations can yield something both coherent and meaningful. The final product can be simple enough and pretty enough for textbooks. But that idealized version may well disguise how it unfolded. The Meselson-Stahl experiment emerged unpredictably from a series of events, sometimes shaped by considerations unrelated to the project's ultimate aim. Yet we can appreciate the opportunistic journey as much as the beauty of the polished product. Great science can emerge from a complex, unpredictable process.

The Dogma of "the" Scientific Method[*]

This is the challenge of science—to shed dogma and get closer to the truth.
—Rudolph Tanzi & Ann Parson, *Decoding Darkness*

It's altogether too easy to reduce all method in science to a simple algorithm. Hypothesize, deduce (or predict), test, evaluate, conclude. It seems like a handy formula for authority. "The" Scientific Method (expressed in this way) haunts the introductions of textbooks, lab report guidelines, and science fair standards. Yet it is a poor model for learning about method in science.

One might endorse instead teaching about the scientist's toolbox. Science draws on a suite of methods, not just one. The methods also include model building, analogy, pattern recognition, induction, blind search and selection, raw data harvesting, computer simulation, experimental tinkering, chance, and (yes) play, among others. The toolbox concept remedies two major problems in the conventional view. First, it credits the substantial work—*scientific* work—in developing concepts or hypotheses. Science is creative. Even to pursue the popular strategy of falsification, one must first have imaginative conjectures. We need to foster such creative thinking skills among students. Second, the toolbox view supports many means for finding evidence—some direct, some indirect, some experimental, some observational, some statistical, some based on controls, some on similarity relationships, some on elaborate thought experiments, and so on. Again, students should be encouraged to think about evidence and argument broadly.

Consider just a few historical examples. First, note Watson and Crick's landmark model of DNA. It was just that: a model. They drew on data already available. They also played with cardboard templates of nucleotide bases.[1] Yes, their hypothesis of semiconservative replication was eventually tested by Meselson and Stahl—*later*. But even that involved enormous *experimental* creativity (essay 4). Consider, too, Mendel's discoveries in inheritance (essay 22). Mendel did not test just seven traits of pea plants, cleverly chosen in advance (as the story is often told). Rather, he seems to have followed *twenty-two* varieties exhibiting *fifteen* traits, hoping for patterns to emerge. He ultimately abandoned those varieties whose results he called confusing.[2] Nobelist Thomas Hunt Morgan, in Mendel's wake, did not discover sex linkage through any formal hypothesis about inheritance. He was looking for

[*] co-authored with Dan Wivagg

species-level mutations in fruit flies. When he first encountered his famous white-eyed mutant, he did not immediately frame a prospective conclusion. Rather, he probed and observed, not sure what he had found.[3] Or consider Darwin. Darwin arrived at natural selection, of course, through synthesizing observations on biogeography, fossils, organismal design, population growth, and limited resources. Only *subsequently* did he reconstruct it as "one long argument" in the *Origin of Species*.[4] In their more recent and monumental work on Darwin's finches, Rosemary and Peter Grant simply extracted significant patterns from voluminous data they collected over many years.[5] No hypothesis. No experiment. No control. If such great heroes of biology did not use the prescribed Scientific Method, how can anyone justifiably portray it as "the" method of science?

Scientific papers do indeed seem to follow the Scientific Method. But they are *reconstructed* accounts of *completed* work. They are composed to fit a standardized *publication* format. They do not describe how research always occurs in practice.[6]

The chief problem may be that through schooling, students come to believe that the Scientific Method (emboldened with capital letters) *guarantees* discovery and unambiguous, reliable conclusions. Uncertainty, incompleteness, and revision are excluded. Of course, science is fallible. But *how* does it fail? The Scientific Method does not say. Henry Bauer has nicely profiled how the mythic method misleads.[7] For Bauer, scientific ideas develop gradually, subjected to successive filters. There is no unique algorithm yielding absolute truth. Students (as future citizens) need to learn how science can be limited, how some evidence can be complex, and how some questions can be unresolved. That, in turn, helps them understand how (or when) we should trust scientific claims. Such judgment is especially important as more and more public decisions involve complex or ongoing science.[8]

Given that the conventional Scientific Method does not adequately describe the richness of science, one may marvel at its hold on the school mindset. Why the entrenched dogma? At one level, the simplicity may be merely convenient. But the dogmas may be deeper. Those who actively defend the Scientific Method seem concerned with the privilege of science. For them, science is special. It is beyond the ordinary. It is exclusive. The Scientific Method demarcates Science with a capital *S*. Without discipline, it would seem, no claim is any better than any other. Order seems secured by rule following and conformism. Portrayed in this way, of course, the promise of the Scientific Method seems grossly overstated. Yet witness how prevalent this perspective is.

Consider instead Albert Einstein's view: "The whole of science is nothing more than a refinement of everyday thinking."[9] Accordingly, a conception of scientific method should grow out of familiar experience. It should complement and extend ordinary discovery processes. And it should highlight how to establish reliable evidence—an aim shared, for example, by journalists and judges. A physician diagnosing an illness, a mechanic troubleshooting a car, a detective tracking a crime all use the same methods as scientists, although in different contexts. The reasoning shouldn't seem foreign. When we apply these same methods to understanding the natural world, we call it science.

If one must characterize method in science concisely, let it be something like this: *Scientists follow hunches, clues, and questions obtained from observations, earlier claims, reading, and so on. They explore how to generate relevant information. They consider possible sources of error. They engage others in interpreting evidence. Results usually lead to more questions. Ideas are refined. Some change; some are abandoned.* Yes, scientists pose hypotheses. Yes, they use controlled experiments. As tools— among others in the scientist's toolbox. Viewing science as constrained by one privileged method is greatly impoverished. We do science in many ways.

PART II

Darwin, Evolution, and Society

6

Was Darwin a Social Darwinist?

Are humans inherently selfish brutes? Skeptics and critics of evolution routinely denounce the ghastly specter of society "red in tooth and claw" as an unacceptable consequence of Darwin's concept of natural selection. They equate Darwinism with so-called Social Darwinism, a belief in ruthless social competition and unmitigated individualism. Many evolutionists, too—even staunch defenders of Darwinism, from Thomas Henry Huxley to Michael Ruse—seem to concur that the natural history of humans leaves an ethical void. Darwin himself, by contrast, had a well-developed interpretation of the evolution of morality.[1] Others since have deepened our biological understanding of human and cultural origins. Perhaps, then, we are ready to challenge this entrenched assumption, this sacred bovine: that belief in evolution entails forsaking any foundation for morality.

Many scientists disavow any role for biology in addressing ethics. They retreat behind the shield of the fact/value distinction or invoke the threat of the fallacy of deriving values from nature. Yet morality is an observable behavior, a biological phenomenon.[2] We might well document it in other species. For example, empathy has recently been observed in both mice and, ironically perhaps, rats. The rats will even forgo chocolate to help a cage mate escape restraint.[3] Morality deserves a biological explanation, especially for those who wonder about the status of humans in an evolutionary context.

There are important limits, of course. One does well to heed philosophers who warn that we cannot justifiably derive *particular* values or moral principles from mere description. Many have tried, and all have failed.[4] "Oughts" do not arise from "ises." Values and facts really do have different foundations. Yet why or how we can express values at all, have moral impulses, and engage in ethical arguments are all psychological or sociological realities, susceptible to analysis and interpretation. Indeed, an understanding of human evolution may well be incomplete without addressing these very important human traits.

Darwin as a Social Organism

One may begin, of course, as one often does with evolutionary topics, by returning to the source: Charles Darwin. How did Darwin regard culture? Did he apply natural selection to society? Was he a "Social Darwinist," as many take his theory to imply?

Well, Darwin had ten children. Was he self-consciously exhibiting awareness that those who reproduce most are ultimately the most fit? If he was, it seems peripheral for those who fret about "survival of the fittest" structuring society. They seem to worry about cutthroat competition for wealth, power, and other social resources. Thus, journalist Robert Wright endeavors to portray Darwin as extremely ambitious, his career replete with "relentless ascent, deftly cloaked in scruples and humility." Wright contends that "He did superbly what human beings are designed to do: manipulate social information to personal advantage." Darwin, he believes, was a savvy political animal: a triumphant "alpha-male" among humans.[5]

Historian and biographer Janet Browne, however, offers a quite different portrait.[6] Darwin was a gentle man as much as a gentleman. He was a loving, even doting father and faithful husband (Figure 6.1). He advocated for the rights of slaves and defended humane treatment for domesticated animals. He wrote explicitly, "if we were intentionally to neglect the weak and helpless, it could only be for a contingent benefit, with a certain and great present evil."[7] In his personal life, Darwin hardly displayed the callousness alleged as inherent in his theories.

FIGURE 6.1 *Darwin as a family man, with his son William in 1842.*

Darwin was also concerned about interpreting human morality scientifically. In July of 1838 he began a private notebook filled with thoughts on metaphysics and naturalistic approaches to mind and morality.[8] (His first thoughts on the transmutation of species were recorded only in May the previous year.) In less than three months, Darwin had filled 156 pages with notes. For example:

> May not moral sense arise from our enlarged capacity or strong instinctive sexual, parental & social instincts, giving rise "do unto others as yourself", "love thy neighbor as thyself". Analyse this out.— bearing in mind many new relations from language.— the social instinct more than mere love.— fear for others acting in unison.— active assistance. &c &c. (*M Notebook*, pp. 150–151)

During that period in 1838 Darwin also read Malthus, the seed that helped him crystallize the concept of natural selection. From the very outset, then, Darwin was thinking about the human and social dimensions of evolution.

In concluding the *Origin of Species* many years later, Darwin advised his readers, with conspicuous understatement, "Light will be thrown on the origin of man and his history." In almost the same breath (although in a now less renowned remark), he also forecast that "Psychology will be based on a new foundation."[9] Darwin knew what he had yet to say.

The Descent of Society

Twelve years passed before Darwin honored his provocative promise. When he did, in *The Descent of Man*, he did not fuss much with the physical body. After all, even Linnaeus a century earlier had comfortably classified humans as primates. (One may thus doubt that objections to human evolution are ever about australopithecine bones, upright posture, or brain size.) Darwin prudently dispensed with the anatomical evidence in a relatively brief opening chapter. His central focus in volume 1 was the emergence of mental powers and morality.

Nor did Darwin address the origin of specific moral principles. Rather, following the philosophical discourse of his day, he focused on moral feelings or sentiments or, as he called it, the "moral sense": crudely, conscience.[10] Darwin's interest in emotion and motivation underscores for us the role of the nervous system. Significantly, for us today, genes are peripheral. Genes may help generate nerves and motivational structures. But nervous systems then operate independently. Memory and learning, not genes, specifically guide open behavior. That same capacity for memory is exhibited by our immune system, which changes when exposed to a particular pathogen, able to respond more quickly to a future infection of the same type. Immunity is a form of learning. Organisms thus learn individually, whether about antigens or external stimuli. Functional flexibility is part of the organism's self-modifying structure. Nervous systems can thereby yield short-term behaviors that function within the longer-term evolutionary filters of survival and reproduction. Understanding a moral sense, as Darwin originally aimed to do, thus draws on primarily psychological, not genetic, explanations.[11] Darwin's own focus on the moral sense thus had great significance for later biologists. He identified where one might find relevant answers.[12]

Darwin postulated four conditions for the emergence of a moral sense. They also reflected prospective stages in its evolution. First, social animals exhibit social instincts of mutual benefit. Second, memory serves as a foundation for conscience. Third, language allows an organism to communicate its needs more effectively. And finally, habit fosters more-immediate responses. Moreover, Darwin asserted dramatically:

> Any animals whatever, endowed with well-marked social instincts, would inevitably acquire a moral sense or conscience, as soon as its intellectual powers had become as well-developed, or nearly as well developed, as in man.[13]

Morality was not just possible, he claimed, but *inevitable* under certain conditions. Here he portrayed evolutionary causality as quite law-like.

Consider each feature more fully. First, Darwin highlighted the role of sociality itself. Since Darwin, of course, the evolution of social organization in other animals has been richly documented. Association with other organisms can be adaptive, even when the individual bears some cost. Once evolved, however, societies may also become a significant further dimension in evolution. Other organisms create a *social* environment. They can shape natural selection, and learned behavior as well. Darwin thus underscored how the values *of the group* would influence individuals. Organisms would thrive socially through "obedience to the wishes and judgement of the community." In a social context, he recognized, self-sacrifice and self-control would be "highly and most justly valued." Eventually, he wrote, "the expressed wishes of the community will have naturally influenced to a large extent the conduct of each member." Darwin recognized variant motives, noting that an individual might ultimately act from "the fear of punishment, and the conviction that in the long run it would be best for his own selfish interests to regard the good of others rather than his own."[14] As examples of such social sanctions Darwin cited macaws screaming disapproval of a mother leaving a nest and baboons slapping a young animal to enforce silence when plundering a garden.[15] In a social environment, natural selection takes quite a different turn. When fitness is partly defined by other members of the group, "survival of the fittest" will tend to promote contributions to the group's welfare, not selfishness alone. Darwin's analysis of sociality thus conforms loosely to the persistent consensus of philosophers about basic moral precepts, such as the golden rule.[16] More deeply, however, by emphasizing sociality Darwin significantly implied that the role of moral assessment emerged at the level of the society or group, not the individual's decisions. "Consequently, man would be greatly influenced by the wishes, approbation, and blame of his fellow-men."[17] Morality, he observed, was intimately related to sociality.

Second, one may consider memory. For Darwin, organisms would encounter conflicts between social and other instincts. Memory would enable retrospective analysis whereby the more enduring social instincts, he imagined, would ultimately prevail. Memory allows integration of short-term and long-term interests. It also allows learning. That would also be critical, for example, to the organism in adopting behaviors that reflected the group's values—either through formal instruction or (as we might say today) through positive and negative reinforcement. With learning, moral education becomes possible, as Darwin implied for the cases of the macaw and baboon. Social values can be instilled and inherited culturally and shape behavior, again quite apart from the modern concepts of genes or instinct.

Third, Darwin gave a role to language. To respond to the needs of others, Darwin noted, organisms needed to be able to interpret their desires, pain, or other mental states. No surprise, then, that the immediate sequel to *The Descent of Man* was *The Expression of the Emotions in Man and Animals*. There, Darwin presented an analysis of body postures and facial expressions, showing how even nonhuman organisms could interpret each other's moods or mental states without words. Even mice seem able to feel another's pain.[18] Morality is rooted in behavior, not just verbal understanding or philosophical argument. The ability to articulate thoughts, nevertheless, clearly deepens the potential for effective interaction.

Fourth, Darwin added habit as important. He saw that some behaviors adopted during an organism's lifetime can become virtually automatic, thereby seeming like instinct. (Even today, ambiguous use of the term *instinctive* confuses the meanings of "innate" and "immediate," or without deliberation.) Darwin imagined, ultimately mistakenly, that habit (namely, repetition) would transform new functions into heritable instinct. In his era, of course, knowledge of learning and inheritance was still limited. Darwin also recognized that the emergence of "social instincts," or cooperation, might be problematic under a framework of individualistic natural selection. He appealed to selection at the level of the group, an idea that continues to be controversial.[19] In both cases, Darwin seems to have stressed instinct and (again) underestimated the potential of learning and cultural transmission of behavior. No doubt he would have been impressed by the later work of Ivan Pavlov, B. F. Skinner, and others, who helped establish just how organisms learn. And not just humans. Not long ago, for example, evidence was reported of explicit teaching in animals: meerkats provide young with live prey to practice prey-handling skills.[20] Chimps also teach younger chimps how to use tools (essay 18). Not all behavior is instinct. Not all inheritance is genetic. Morality need not be "in our genes" (see essays 9 and 19). Through learning, the moral sense, as Darwin suggested, might ultimately come to be habit, or "second nature."

Reconceptualizing Darwinism

Darwin had a well-developed theory of a moral society, then, even if incomplete and, by today's standards, needing revision (for updates, see essay 9). He hardly endorsed the dog-eat-dog world that many contend his theories imply. Quite the opposite. He profiled how sociality and a social environment would limit and counterbalance any individualism. Darwin was no Social Darwinist. Indeed, one may wonder how such a name became affixed to so un-Darwinian a perspective—a puzzle addressed in the next essay.

Darwin is widely known for addressing what he once called "that great mystery of mysteries" or, as expressed in his famous title, "the origin of species." Yet Darwin also provided valuable benchmarks for answering another great "mystery of mysteries": how a moral society might originate through natural selection. Contrary to widespread assumptions, evolution does not entail an ethical void. Darwin indeed helped throw light on our moral heritage, opening a field of inquiry that yields ever-deeper insights as it continues.[21]

7

Social Un-Darwinism

It is time to rescue Darwinism from the dismal shadow of Social Darwinism. According to this now widely discredited doctrine, human society is governed by "the survival of the fittest." Competition reigns unchecked. Individualism erodes any effort to cooperate. Ethics and morality become irrelevant. Some contend that social competition is the very engine of human "progress," and hence any effort to regulate it cannot be justified. Others accept competition as inevitable, even though they do not like it or endorse it. They seem persuaded that we cannot escape its presumed reality. Natural selection, many reason, is ... well, "natural." Natural, hence inevitable: what recourse could humans possibly have against the laws of nature? Thus even people from divergent backgrounds seem to agree that this view of society unavoidably follows from evolution. Creationists, not surprisingly, parade it as reason to reject Darwinism outright.[1] By contrast, as resolute an evolutionist as Thomas Henry Huxley, "Darwin's bulldog," invoked similar implications even while he urged his audience to transcend them morally.[2] Yet the core assumptions of so-called Social Darwinism are unwarranted. Why does it continue to haunt us? The time has come to dislodge this entrenched belief, this sacred bovine: that nature somehow dictates a fundamentally individualistic and competitive society.

Unraveling the flawed argument behind Social Darwinism also yields a more general and much more important lesson about the nature of science. The historical argument seems to enlist science to portray certain cultural perspectives as "facts" of nature. *Naturalizing* cultural ideas in this way is all too easy. Cultural contexts seem to remain invisible to those within the culture itself, sometimes scientists too. The case of Social Darwinism—not Darwinism at all—illustrates vividly how appeals to science can go awry. We might thus learn how to notice, and to remedy or guard against, such errors in other cases.

Social Darwinism without Darwin

Ironically, the basic doctrine now labeled "Social Darwinism" did not originate with Darwin at all. Darwin was no Social Darwinist. Quite the contrary: Darwin opened the way for understanding how a moral society can evolve (essay 6). Indeed, by Darwin's era, the notion of unregulated selfishness as a "natural" condition that threatened social order was centuries old.

In the mid-1600s, for example, Thomas Hobbes described the primitive state of nature as "*bellum omnium contra omnes*": a war of each against all. For him, supreme individualism (if left unchecked) would eclipse sociality. Even genuine benevolence seemed impossible. In Hobbes's cynical spin, generosity was really disguised self-interest: "For no man giveth but with intention of good to himself; because gift is voluntary; and of all voluntary acts the object is to every man his own good."[3] Hobbes's proposed solution was to imagine a social contract. If everyone agreed mutually to limit self-serving behavior, all would benefit. *If.* Anyone could cheat. As in a legal system, who enforced the contract? One would need a moral authority outside or above the system (for Hobbes, it was the king). The lack of moral grounding—evident in the dilemma of cheaters and the absence of warrant for authority—was the same that critics of evolution now fault in Social Darwinism. And it resulted from the same basic assumptions: individualism and the "war of nature"—both posited without (and well before) Darwin.

"Social Darwinist" perspectives were also expressed by Thomas Malthus in his 1798 *Essay on the Principle of Population* and in its many subsequent editions. For Malthus, population would forever increase ahead of the ability to feed it. The "natural inequality" of population and production, he claimed, spoke against romantic ideals (then prevalent) of social improvement:

> Necessity, that imperious all pervading law of nature, restrains them [the seeds of life] within the prescribed bounds. The race of plants and the race of animals shrink under this great restrictive law. And the race of man cannot, by any efforts of reason, escape from it. Among plants and animals its effects are waste of seed, sickness, and premature death. Among mankind, misery and vice.

> No fancied equality, no agrarian regulations in their utmost extent, could remove the pressure of it even for a single century. And it appears, therefore, to be decisive against the possible existence of a society, all the members of which, should live in ease, happiness, and comparative leisure; and feel no anxiety about providing the means of subsistence for themselves and [their] families.[4]

For Malthus, limited resources led to competition. A "struggle for existence," as he phrased it, was inescapable. Both Malthus and Hobbes hoped for a solution, of course. But they could do little more than appeal to awareness and self-restraint, at odds with their very assumptions. Assumptions of individualism lead to conclusions of individualism, no surprise.

Malthus also went significantly further. He viewed efforts to alleviate poverty as only further compounding the problem of population. He thus recommended abandoning the Poor Laws as useless welfare. (He did not explain why, meanwhile, the wealthy should keep their ease and "comparative leisure.") Here, the "war of nature" had acquired a new ideological stance. Idealized values now permeated observed facts—but only by sidestepping the customary need for moral justification. Malthus's cultural values became disguised as a scientific conclusion.

In all this, Darwin was nowhere yet present. Social Darwinism thus seems grossly misnamed. Views about inherent social competition and humans as subject to some "war of nature" should be called, more appropriately, *Hobbism* or *Malthusianism*.

The possibility of Social Darwinism without Darwin only amplifies the puzzle of why Darwin's name should be associated with the doctrine. One may begin to see that Social Darwinism is not a social application of Darwinism so much as an effort to "Darwinize" a social ideology. That is, the claims about society were never derived from science. Rather, proponents of a particular cultural perspective appealed to Darwin and science in seeking authority for their views.

Appropriating Nature through Darwin

The introduction of Darwin's (and Alfred Wallace's) concept of natural selection did indeed significantly alter political discourse—but not because there were suddenly new ideas about humans or society. Rather, what changed was how the already existing ideas were justified. Those who believed in Hobbism or Malthusianism readily interpreted Darwin's new perspective as further exemplifying their views. They appealed to crude analogies and Darwinian-type arguments to argue that competition was both "natural" and "progressive." Natural improvement, they implied, trumped any other moral arguments. That is, they endeavored to *naturalize* their social doctrine. This pattern of reasoning, shared by a cluster of thinkers in the late nineteenth and early twentieth centuries, later prompted historian Richard Hofstadter to coin the very term "Social Darwinism"—nearly an entire century *after* Darwin's great work.[5]

Hofstadter's "Social Darwinists" were misled in two primary, related ways. First, their analogy from nature to society did not, ultimately, follow Darwin's arguments at all. The socialized version of "survival of the fittest" was biologically flawed. Second, they mistakenly viewed adaptation to local environmental conditions as progress, or improvement, on some imagined scale of abstract value. They erred scientifically—not just by layering values on facts, as commonly noted. The unwary can thereby easily succumb to the same mistakes, unless they fully understand how the errors arise.

The first error was the looseness with which some applied Darwinian principles to society. They did not fully consider the structure of selection in social environments. The relationship seems simple enough. After all, Darwin and Wallace had both capitalized on an analogy *from society to nature* in first conceiving natural selection. Their insights were each triggered by reading Malthus and applying his social "struggle for existence" to organisms. For them, however, it was a guiding analogy only: a template or scaffolding soon girded with biological principles and observations.[6]

By contrast, the reverse analogy of the Social Darwinists, *from nature to society*, fails. For example, political power and economic class are not biologically heritable traits. Social stratification is not the result of differential survival or differential reproduction, even metaphorically. Social privilege is not genetically based. Nor is it even clear *what* is selected. Moreover, humans actively shape the environment. Social "selection" is a diffuse amalgam of human choices. In the casual social analogy, therefore, neither the heritable units, nor the sources of variation, nor the units of selection, nor selection itself fits Darwinian patterns. Natural selection does not map onto social politics.

A closer analogy, if any, might be Darwin's *artificial* selection, the concept that farmers deliberately and selectively breed certain desirable types. But that clearly strips the analogy of its "natural" context. Social "selection" is not fixed by circumstances beyond our control. We choose. Sociality, as Darwin hinted in *The Descent of Man*, introduces a distinct set of dynamics based on behavior and information transfer. Cultural evolution follows its own principles, shaped by learning and the activity of minds. The Social Darwinists' concept that some individual humans are objectively fit while the poor or indigent are unfit is utterly misframed. The whole analogy—and the ideological justification that relies on it—collapses. The historical Social Darwinism was *scientifically* misguided and unwarranted. No wonder perhaps that the major advocates of the doctrine were not scientists, but a social elite who would benefit if the purported "science" legitimated their wealth and power.

The second critical error of Social Darwinist thinking was subtly inscribing a human value into nature. Adaptation through natural selection seemed, through human eyes, progressive. And who could argue against progress? As biologists well know, however, evolution is not uniformly progressive on some absolute scale. Context is essential. "Fit" organisms thus vary from location to location and from time to time. Competition in nature does not lead to a generalized improvement, measurable abstractly. In the social context, arguments about competition obscure the significance of particular contexts and consequences. Amidst the confusion, a vague and ill-defined progress seems like a trump card to eclipse the relevant arguments. The view that Darwinian selection itself was inherently progressive was not independent of the cultural value of that concept.

Confusion may also arise in part from anthropomorphizing the term "selection." The word seems to imply that natural selection is a choice, rather than the effect of nonintentional causes. Survival and reproduction are not values, any more than falling acorns express a value of gravity, or oppositely charged ions a value of electrical attraction. Intention and moral agency come ultimately from minds. As flexible adapted structures, brains can develop and recognize many values. The process of evolution itself, however, does not exhibit intent or values, even if certain organisms are "selected."

The Naturalizing Error

Both Social Darwinist errors appeared in the work of self-made philosopher and social commentator Herbert Spencer (Figure 7.1).[7] Spencer presented evolution as

FIGURE 7.1 *Herbert Spencer, originator of the phrase "survival of the fittest" and major promoter of applying evolutionary ideas to society.*

due to a natural unfolding from simple to complex. Chaos and sameness developed into integration and diversity. Hence, he used—and popularized—the misleading term "evolution," which means, literally, unrolling. Spencer's evolution, notably unlike Darwin's descent with modification, was inherently progressive. Further, Spencer appealed to that view to defend a "hands-off" social policy. Abolish the Poor Laws, he urged, because they interfere with "natural" progress.

Spencer also blurred the organic/social distinction. Using one grand unified scheme, he espoused continuity from anatomical adaptation to individual mental development to the history of societies and their institutions. For him, society *was* nature, merely at a more "advanced" stage.

Spencer's universal principle of unfolding progress and his biologizing of society inspired many others to condemn social welfare and, especially in America, to argue that capitalism ought not to be regulated. Today, Spencer's ideas lie in disrepute. But his historical influence was immense. His works were read far more widely than Darwin's. Over six hundred thousand copies of his books were sold. Hofstadter's "Social Darwinists" were thus largely *Spencerists*. Accordingly, the effort to naturalize a laissez-faire social policy might well be called *Spencerism*.

Spencer's work was eventually criticized by philosopher G. E. Moore.[8] Moore faulted Spencer's appeal to natural events and processes as an implicit model. Nature was not a source of values. The reasoning was *ethically* unwarranted. Moore called it "the naturalistic fallacy." Moore did not really address the biology. For all he cared, the claims about a naturalized society might nonetheless be *descriptively* warranted. He thus mistook the principal error. Spencerism was not just a lapse of *moral* reasoning. Rather, it was faulty *scientific* reasoning. Spencer was trying to inscribe cultural views into nature—what one may call, instead, *the naturalizing error* (also see essays 16–20).[9] Nowadays, few persons endorse Spencerism, but many imagine it to be true. They escape Moore's naturalistic fallacy in ethics but succumb to the naturalizing error in science. Spencer's social analogies were *biologically* unwarranted. Adjusting modern popular beliefs thus involves, as an initial step, nurturing a proper biological understanding of the origins of society and culture (essay 6).

Given the risk of the naturalizing error, how does one maintain the integrity of science? One might want to isolate facts and values and never let them mix. Yet scientific findings frequently inform our judgments. We disregard the relevance of facts to values at our cost. The challenge, rather, is to focus on the justification. "Follow the values." Be alert to the cultural contexts of "facts." Tracing the source of the values, for example, may help decide whether "scientific" claims are responsible or reflect cultural bias. In the case of evolution and society, the pattern was clear to Friedrich Engels, who commented on them in an 1875 letter:

> The whole Darwinist teaching of the struggle for existence is simply a transference from society to living nature of Hobbes's doctrine of *bellum omnium contra omnes* and of the bourgeois-economic doctrine of competition together with Malthus's theory of population. When this conjurer's trick had been performed . . . the same theories are transferred back again from organic nature into history and it is now claimed that their validity as eternal *laws* of human society has been proved. The puerility of this procedure is so obvious that not a word need be said about it.[10]

"Not a word" may have been needed for Engels' correspondent. But for those unfamiliar with the history and nature of science, a full explanation is surely much needed, if not essential. The Social Darwinists drew inappropriately on the authority of science. For them, Darwin was nothing more than a convenient form of social justification. They were never scientific. And it is the masquerade as science that matters most deeply.

Guarding the Integrity of Science

There is nothing Darwinian about Social Darwinism. Darwin did not endorse it. The beliefs can—and did—exist without Darwin. The pattern of thinking is not Darwinian, although it borrows its terminology. There might be a Darwinian approach to society—indeed, as sketched by Darwin himself—but not one based on an ideological framework.

"Social Darwinism" (as commonly understood) is a pernicious misnomer. The very name defames science, especially Darwinian concepts, by portraying an ill-informed cultural interpretation of science as an extension of science itself. We should purge the phrase from our language. We should challenge it as inappropriate everywhere and every time it is mentioned. We should, instead, talk about the social doctrines of Hobbes, Malthus, and Spencer—and the dangers of the naturalizing error exemplified in their views.

Spencerian errors, of course, persist in our time. Why? The most significant contributor may be another ostensibly Darwinian phrase, "survival of the fittest." A mere four words seem to embody the whole naturalizing analogy, apparently describing nature and society equally and guaranteeing the best outcome. Interpreting this prospective nemesis is yet another important challenge, addressed in the next essay.

8

A More Fitting Analogy

In our culture no one needs a biology class to learn about "survival of the fittest." Yet one might need instruction to unlearn the misconceptions engendered by the analogy's potent imagery.

In popular perspectives, this single phrase conjures images of humans—however civilized—as brutish organisms vying for jobs, status, and power. Maybe they also compete for prime mates. The language of "survival" resonates with "survivor" contests on television: "Outwit. Outplay. Outlast." Mention of the "fittest" implies that physical "fitness" and athletic prowess are ideals. At the same time, the extremeness of reference to only the "fittest" implies that a human's fate seems to be life *versus* death, fit *versus* unfit, winning *versus* losing. In all, cooperation and coexistence give way to warfare, conflict, and backstabbing gossip: "society, red in tooth and claw," to adapt Tennyson's phrase. Mostly, life reduces to *competition*. Cutthroat competition. Through just one expression, all these interpretations seem to have a biological basis. None of them are scientifically justified. "Survival of the fittest" is not a neutral phrase, idly describing natural selection. Instead, through unintended metaphors the language fosters major misconceptions.

One might hope to remedy these many confusions. But how? "Survival of the fittest" seems to describe both organic evolution and human culture. So first, one must carefully distinguish the processes of each, functioning at different levels (essay 6). Second, one needs to understand how ideology can be unduly naturalized (or improperly inscribed in "nature"). The phrase was never purely descriptive. It expressed cultural values (essay 7). Even with both these pitfalls resolved, however, problems may persist because of the very language itself. The connotations of the phrase seem inescapable. Here, I consider the misconceptions latent in each individual term: "survival" and "fit"—as well as the "-est" suffix. That may help us craft a more fitting analogy or catchphrase to describe natural selection.

Surviving

Consider first the implications of the word "survival." What matters to evolution is differences in survival rate at a population level. Differential survival leads to differential reproduction, the essence of natural selection. For example, Darwin talked of "proportional numbers."[1] Yet the phrase "survival of the fittest" invites us to think starkly in terms of individuals. The outcome seems reduced to survival or death of just one organism. Fit organisms live; unfit ones die. In the "struggle for existence," one either succeeds or fails. Selection becomes either-or.

When biologists use this expression, they unwittingly help perpetuate a culture that tends to acknowledge only winners and losers, survivors and also-rans. The language of crude "survival" distorts the biological concept and promotes inappropriate cultural overtones, as well. Shifting from a world of black and white to a world of nuanced grays is, of course, part of growing up. But biological textbooks often highlight predation as black and white, fostering misleading impressions that persist well past the biology classroom (see essay 20).

As an alternative, Darwin called natural selection a "principle of preservation": namely, the "preservation of favourable variations."[2] He referred to "the strong principle of inheritance" and how "any selected variety will tend to propagate its new and modified form."[3] Darwin thereby underscored the importance of continuity and propagation as integral to selection. A focus on survival, by contrast, lessens the significance of reproduction, successive generations, and the long term. Death seems primarily to weed out the unfit. Selection seems to act negatively, as a screen or filter. It becomes *eliminative*.

This image is echoed in our culture, even in how we elect to entertain ourselves. In television game shows, "survivor" contests, beauty pageants, and talent searches, contestants are successively eliminated. They are often pared down one by one. In sports—from softball to drag racing to tennis—one frequently finds *double elimination* tournaments. Athletic championships are often decided not by overall season performance but through stepwise reduction in end-of-season play. Why do we choose to enjoy ourselves this way? Why is elimination the commonly accepted framework? Are we indoctrinated to regard it as "natural"?

With a primary focus on survival and elimination, it is hard to appreciate how natural selection can be creative or generative. A focus on reproduction, by contrast, underscores how the process is repeated over and over again. Fitter variants proliferate. Adaptive organisms flourish. Selection leads to expansion, even without reduction. Favorable traits, once preserved, can accumulate. New forms can emerge. Highlighting *amplification* instead of elimination changes the whole image of the process. Creation replaces destruction. Diversification replaces extinction. Growth and development replace death. One might characterize natural selection more positively, then, as differential amplification: "the fecundity of the fittest" or "the flourishing of the fit."

Being Fit

Consider next how we commonly perceive fitness, or being "fit." In Darwin's time, fitness already had a nonbiological meaning: appropriateness, or "answering the purpose." Today we still might say "an action fit for the occasion" or "a fitting analogy." (Like well-fitting gloves or jeans?) Darwin's contemporaries would have easily understood the new biological meaning: an organism's being well suited to its natural environment. Darwin invited his readers to note "how infinitely complex and close-fitting are the mutual relations of all organic beings to each other and to their physical conditions of life."[4] Organisms adapt *to* something. Natural selection occurs *with respect to a local environment*. Fitness develops meaning only *in context*.

Since Darwin's time, the meaning of fitness outside biology has shifted. One significant development (especially in America) was the physical fitness movement and the introduction of physical education into schools. Now, athletics permeate popular culture and designer fashions, business networking, perceptions of college prestige, and international Olympic stature. Sports earn their own section in daily newspapers and exclusive channels on television. Our culture today regards fitness primarily as *physical* fitness. And how esteemed it is! Advantage seems to depend on being stronger, fleeter, hardier, more agile. Selection seems primarily *athletic*.

Darwin himself referred to the "struggle for existence" and the "battle for life," sometimes even drawing on militaristic metaphors (such as defense, shields, attacks, or the "war" of nature). Yet such expressions were also balanced by other images. For example, after describing the swiftest and slimmest wolves as best able to capture prey, Darwin discussed nectar-laden flowers as best able to attract insect pollinators. Nectar-gathering insects, likewise, benefitted from "curvature or length of the probiscis."[5] Darwin's finches, of course, provide a classic example, modestly based on foraging, not fighting. Fitness is based on functional context. Ideally, one views fitness primarily in terms of efficient foraging or resource use, say, rather than athletic ability.

Athletic images of fitness also further confound understanding of the reproductive dimension of natural selection and sexual selection. Of course, one may easily succumb to a caricature of athletes flexing muscles for sexual stature among prospective mates. Darwin himself described how male chickens used their spurs in competing for females. Yet he subsequently cited examples "of a more peaceful character": birds that secured mates by being "the most melodious or beautiful"[6] (Figure 8.1). How different natural selection seems when exemplified by warblers, orioles, toucans, or birds of paradise, rather than lions and tigers and bears! In *The Descent of Man*, Darwin described at length the role of gaudy plumage, noisy insect calls, and showy antlers in attracting mates. Other traits enhance reproductive success as well, such as large-yolked eggs, internal fertilization, internal development, nesting, and feeding and teaching offspring. Is it beyond our ability to make the generation and nurturing of offspring as dramatic or apparently compelling as the athletic competition of the World Series or NASCAR racing?

Natural selection may also be well illustrated in how our bodies' immune systems respond to infection. All the essential concepts of selection are exhibited here. First, a large reservoir of different, specific immune cells provides *blind variation*. Second, a uniquely shaped foreign body, or antigen, in the blood provides the "*environment*." *Fit* proteins (immunoglobulins) from the white blood cells are those that literally "fit" the antigen: the shapes are complementary. Consequently, the well-adapted cells respond and *proliferate*. The potential to recognize the pathogen is

FIGURE 8.1 *The peacock's showy tail, a familiar example of the significant role of differential reproduction in evolution.*

amplified. Biologists call this "clonal selection." It is a valuable analogy for understanding the central mechanism of natural selection without engaging all the cultural imagery (and politics) of evolution.

A biological focus on resources and preservation, rather than athletic or militaristic traits, begins to indicate an alternative characterization. A modern echo of the old term "fit" would be "apt." How fortunate, perhaps, that "apt" is the very root of "adaptation" and "adapted."

Being the "-est"

Consider, finally, what at first may hardly seem worth the notice: the "-est" in "fittest." The superlative suffix, far from being insignificant, accentuates the comparison between organisms. "Fit" alone no longer suffices. One must be *more* fit. Selection becomes fundamentally *competitive*. One must also be more fit than any other. One must be the *most* fit. Selection becomes *winner-take-all*. With such apparently high stakes, no wonder perhaps that selection also seems primarily *antagonistic*.

Competition surely featured in Darwin's own conception of natural selection. Darwin, like many others at the time, saw nature as luxuriant and full. New species, he imagined, would have to "wedge" their way in among those already filling available niches and using available resources.[7] Darwin also applied a Malthusian perspective about society to biology: scarcities would lead to a "struggle for existence." Indeed, Victorian culture viewed both culture and nature as fundamentally harsh. Envision Charles Dickens's London: poverty, slums, child labor, and grim working conditions. Part of Darwin's triumph was to conceive how even from

Malthusian-like conditions, adaptive design might nevertheless develop. "From the war of nature, from famine and death," he rhapsodized in his conclusion, " . . . endless forms most beautiful and most wonderful have been, and are being, evolved."[8] Darwin adopted, yet also grandly transformed, the competitive views of his culture (essay 12).

Our culture is not Darwin's. Yet competitive rhetoric still reigns. For example, nations, cities, and businesses strive for "economic competitiveness": an abstract ability to compete seems to upstage the goal of ultimate economic sustainability or well-being. Once, an economy was a system for exchanging goods and services; now, it's seen as market forces. Once, education was a forum for knowledge and understanding; now it's about being able to "compete" for jobs. (And in the short term, of course, it's all about competing for grades and admission to the best schools.) Even recreation frequently drifts into competitive games, as though we could not entertain ourselves without winners and losers. Cheerleading, once about rallying competitive spirit among others, is now competitive itself. Somehow, the culture succumbs to Spencerian doctrines that unfettered contests lead to a better world (essay 7). Would such competitive frameworks be so pervasive if the "survival of the fittest" language did not implicitly render them as "natural" and progressive?

Moreover, social competition is often seen as winner-take-all—reflected in a widely parroted creed that "*only the fittest* survive." How many seem to care about second best, whether in the Super Bowl or some talent contest? Even democratic governance seems reduced to "winning" just enough support to eclipse the "opposition." Elections, too, seem less systems of collective discourse and choice than mudslinging competitions for votes. Winning and excluding seem more important than developing a community-wide consensus. Anyone challenging these norms typically gets an earful of mangled natural selection concepts blindly applied to social contexts. The very fabric of American society—from economy and governance to education and recreation—seems permeated with the "survival of the fittest" theme of winner-take-all competition.

Natural selection may surely be propelled at times by competition. Yet selection also occurs widely without it. For example, the first organisms to venture on land flourished more by escaping competition than by "beating" it. Entering new adaptive zones, proliferating, and diversifying is a familiar evolutionary theme. Indeed, that would be the story of the first finch from the South American mainland to reach the untapped Galápagos archipelago, whose lineage later diversified into the famous Darwin's finches. Selection can reflect capitalizing on opportunity. In other cases, organisms adapt by tolerating stressful habitats—low in water, nutrients, light, or other vital resources, or at extreme temperatures, pH, and so forth.[9] In yet other cases, in frequently disrupted environments, organisms adapt by being "ahead of the competition." They exploit the potential to reproduce and disperse rapidly, rather than compete directly. Life strategies vary widely. Competition for limited resources is not the only pattern.

One may also want to remedy the perception of selection as crudely selfish, and hence antagonistic. Many organisms thrive through cooperation with other species. Mutualisms abound. Pollination (Figure 8.2) and seed dispersal symbioses) are

FIGURE 8.2 *Pollination mutualism between flower and insect: enhancing survival through cooperation, not physical contests.*

widely known, but perhaps too often relegated to the shadows. Another remarkable set of mutualisms, in the news recently, involves sea slugs. As unshelled mollusks, sea slugs are soft and vulnerable, and move, well, at a snail's pace. They are quintessentially *un*athletic. Yet by hosting algae or chloroplasts in their digestive glands, some can survive for months without food.[10] In this case, selection has amplified cooperative abilities. Sociality can evolve too. Mutualisms may be within species, as well. Ironically, the presence of cooperative behavior hardly seems to stem the common competitive mindset. That may bear witness to the potency of the "survival of the fittest" metaphor.

An Alternative Darwinian Gestalt

In summary, in our culture, the phrase "survival of the fittest" is misleading biologically while lending ghastly inappropriate support to many social metaphors. We should abandon this popular phrase (another sacred bovine?). Biologists especially should explicitly expose its pretensions.

But let us not pretend that we can proceed without an alternative sound bite. We need a succinct, memorable substitute. We need a catchphrase that does not frame selection as inherently (1) eliminative, (2) athletic, or (3) antagonistic. Selection is not and need not be framed as (4) either-or, winner-take-all competition. Instead, an ideal alternative will highlight (1) the role of reproduction and the flourishing of lineages, (2) context and local resources, (3) multiple potential life strategies, and (4) opportunism. While we're at it, why not find something a little less severe? Without romanticizing nature, can we portray a brighter, more congenial Darwinism?

Perhaps we might take a cue from Darwin himself. In closing the *Origin of Species*, he invoked the image of "an entangled bank, clothed with many plants of many kinds, with birds singing on the bushes, with various insects flitting about, and with worms crawling through the damp earth." He found "grandeur in this view of life": rich, diverse and fertile. The image echoed thoughts Darwin recorded decades earlier in his private notebooks, where he drew ostensibly the first evolutionary tree. Some lineages expand. They diversify. As in living trees, when one branch dies, other branches grow and fill the opportune gap. The core image is one of flourishing. And organisms flourish because they aptly fit their context. Hence, they generate, as Darwin penned so eloquently, "elaborately constructed forms, so different from each other, and dependent on each other in so complex a manner."[11]

Here, then, is a prospective alternative to "survival of the fittest": *amplification of the adapted*. That is, one may rephrase Darwin's own definition as follows:

> We may call the principle, by which variant organisms well suited to their environment are preserved and flourish, Natural Selection, or Amplification of the Adapted.

That may not be perfect. But it may suffice.

Or do we need a prize competition to find the best expression?

9

The Domesticated Gene

Enough of the "selfish gene," already. It was a clever mental game, once, to imagine that genes are the ultimate units of evolution. That organisms are no more than a gene's way of making more genes. The concept fundamentally confuses levels in biology. It implies that genes can have intent and moral perspective.[1] The anthropomorphism is grossly misleading. Even Richard Dawkins, who originally launched the concept, now seems to acknowledge as much.[2] When the notion was introduced, sociobiology was also new, promoting an evolutionary and genetic view of behavior. It was all too easy to consider *all* behavior, like genes, as "selfish." Explanations of cooperation, "altruism," and social reproductive behavior were reduced to genes through the concepts of inclusive fitness and kin selection. The supremacy of the individual seemed to epitomize Darwinism. These perspectives thus gradually became entrenched, and now appear as fact in virtually every textbook: another sacred bovine?

Recently, however, E. O. Wilson, the prominent founder and advocate of sociobiology, has renounced kin selection in explaining societies with a single reproductive individual.[3] The idea was that genetic relatedness could explain why some individuals did not themselves reproduce but instead helped others, their kin, reproduce similar genes. Three decades of research have shown that many cooperative breeding societies (such as termites) do not exhibit the required genetic structure of haplodiploidy. Moreover, many species that do (including sawflies and horntails) are not social. The documented cases and the explanation do not align. Rather, the societies in question—from ants and honeybees to beetles, shrimp, and naked mole rats—all seem to have nests with restricted access, guarded by just a few individuals. The social cooperation seems just an "ordinary" adaptation to certain conditions. The striking reproductive structure, Wilson now contends, is an evolutionary consequence—not a cause—of the social organization.

Wilson's dramatic turnabout illustrates a wider shift in perspective.[4] For decades William Hamilton's notion of kin selection largely eclipsed Robert Trivers's concept of reciprocity as an explanation for cooperation.[5] While the former is commonly presented in educational contexts, the latter has been nearly always absent.[6] But

the once-popular reductionism is now yielding to social and cultural evolutionary approaches. In case after case, reciprocal interactions—immediate or deferred,
direct or indirect, and mediated by rewards, sanctions, or reputation—have
emerged as significant. As Wilson's claims suggest, higher levels of organization
can govern lower-level genetics. The "selfish" gene has been *domesticated*. What are
all the new findings that have dramatically reoriented our view of the evolution of
cooperation?[7]

When Nonkin Cooperate

The explanatory limits of kin selection are plainly evident where nonkin cooperate.
Why help organisms that do not carry on your genes? Consider, for example, the
case of two unrelated capuchin monkeys, Sammy and Bias, described by primatolo-
gist Frans de Waal.[8] They had learned to jointly pull a spring-loaded tray to obtain
food: a simple form of cooperation. On one occasion, however, Bias did not get
her share before the tray snapped back out of reach. She threw a tantrum. Sammy
finally returned to the task, and Bias, too, got her reward. Both clearly understood
the implicit contract of mutual effort. Quid pro quo. Following Trivers's model,
their help was based on reciprocity. In the same way, de Waal observes, unrelated
chimps will share meat after a hunt. Often, they repay social favors. Indeed, in sustained observations of a captive chimp troop, the exchange of grooming and special
food items tended to even out in the long run.[9] Sharing was not based on kinship.
But neither was it blind.

Recently, Cambridge zoologist Tim Clutton-Brock reviewed the growing list of
cases of cooperation between nonkin in animal societies.[10] For example, stickleback
fish share risk in approaching predators (such as trout) to assess the degree of threat.
Olive baboons assist each other in mating competition. Hooved animals exchange
grooming. Not only do rats cooperate, but their tendency to do so grows with experience. One may add all sorts of cases of cooperative predation (such as among peli-
cans), as well as defense against predators (such as mobbing by meerkats). In such
cases, the mutual benefits are immediate. The consequences for natural selection are
easy to imagine. They echo the mutualisms between species familiar from biology
textbooks and nature television. But these should be distinguished, Clutton-Brock
cautions, from occasions where the benefits are deferred—more problematic in an
evolutionary context.

The hitch is that helping behavior may never be reciprocated. Delay opens the
way to free-riders: organisms that harvest the benefits while contributing nothing.
Reciprocity, when deferred, requires trust or long-term accounting. The problem
of lag time was investigated with blue jays by David Stephens at the University of
Minnesota.[11] His lab examined several rounds of potentially cooperative interactions, but added a key feature: delaying the payoffs. The birds could observe each
other's behavior and assess it after each step. The delay allowed trust to develop
before the fruits of sharing became available. Blue jays will extend cooperation and achieve greater rewards, the researchers found, when time is allowed for

accountability. Kinship aside, cooperation can occur through reciprocity when trust can be established (or enforced—see below).

The kin selection bandwagon was thus premature. For example, Florida scrub jays stay at home and help raise their siblings—even when they are reproductively mature. That behavior was quickly labeled an example of kin selection. However, Glen Woolfenden and John Fitzpatrick showed how it was a social adaptation. As a result of helping, males secure better territory. They work to enlarge their father's range and then cleave off their own. This is easier than trying to wedge a new territory into space that is already claimed. Females, too, by "sitting and waiting," secure better mates. They can be choosier. Individual benefit, not kinship, shapes scrub jay cooperative behavior.[12]

Accordingly, one must question the role of kin selection even where organisms are related. Indeed, mathematically, the conditions for kin selection prove quite narrow: it can occur only when selection is weak and benefits are strictly additive. For example, selection cannot depend on the density of other organisms and the frequency of beneficial exchanges. Yet when cooperators become more common, cheaters benefit more. Also, fitness depends on the whole community, not just immediate neighbors. One research team has corrected the original formula for calculating kinship benefits by taking population structure into account as another variable. In the cases they studied—microbes—those populational factors were far more significant than genetic relatedness.[13] E. O. Wilson bolstered his recent criticism of kin selection with similar mathematical analyses by two colleagues.[14] They highlighted the unrealistic requirement in the original view that all interactions be directly one on one. Nature is more complex, they claim.

Ultimately, kin selection may be relatively rare. Inclusive fitness, important in interpreting the evolution of sex ratios and certain reproductive trade-offs, seems limited in explaining cooperation and sociality.

When Society Regulates Individuals

Because delayed reciprocity is susceptible to cheaters, cooperation cannot evolve if they proliferate. But solutions may emerge through social interactions and individual choice (not at the genetic level). De Waal modeled the basic dilemma in one experiment with his capuchins. After the monkeys had learned a basic cooperative drill, he restricted the food reward to only one individual. The privileged monkey (playfully dubbed the "CEO") usually shared the prize. When he did not, however, on subsequent trials the would-be cooperator went "on strike."[15] Here, failure to share was kept in check by social accountability. "Opting out" was a modest, local form of peer pressure.

The same pattern is reflected in vampire bats (Figure 9.1). The bats share blood, accommodating the risk of not feeding on any particular night. But repeat beggars are denied help if they have not reciprocated.[16] Again, cooperative behavior is enforced by social interaction, not genes. Darwin expressed it a bit more eloquently, of course: a human, he wrote, tends to be "greatly influenced by the wishes,

FIGURE 9.1 *Vampire bats enforce a system of cooperation for sharing blood by punishing cheaters.*

approbation, and blame of his fellow-men."[17] Either way, negative social consequences seem to shape individual behavior.

Yet sanctions are costly to the enforcer. Are such costs balanced by individual benefits? Is the system sustainable? In many mathematical models, a system where noncooperators are punished seems stable, once established. However, where cheating is already prevalent, individual punishers are ineffective by themselves. So a recent model considered *coordination* of punishment. The ability to gang up on free-riders proved critical in creating cooperative groups.[18] That finding seemed to resonate with earlier anthropological claims that such coordination helped early humans to level primate hierarchies, yielding egalitarian societies.[19] Newly emerged language may have facilitated that coordination.[20] Another possible mechanism, demonstrated in other models, is allowing individuals to choose their group. Groups that punish cheaters flourish at the expense of unregulated groups.[21] Both strategies yield a social system of cooperation enforced through punishment.

One need not limit effective interactions to punishment, however. Rewards also work. Indeed, in one study of almost two hundred human subjects, rewards became more common as the group interacted (while punishment waned). The outcome was higher levels of sharing.[22] Incentives can be as effective as sanctions.

Another study considered the relative roles of both rewards and sanctions. Participants were sorted into short-term pairs in successive rounds. One was designated a donor, the other a recipient. The key variable was that the donor received information about the recipient's recent behavior when acting as a donor in earlier rounds. The donor then chose to: (1) give the other a donation; (2) give nothing, or (3) impose a cost on the recipient. Because there was no symmetrical interaction, one could not enforce reciprocity directly. The effects of all choices were indirect. Ultimately, generous givers fared better as recipients than free-riders. In the long run, punishment was minimal. Still, based on comparison with a control group, the ability to punish critically kept free-riding in check. Incentives and sanctions in tandem helped generate a system of indirect, or network, reciprocity.[23] That could explain why even in a large society, persons might incur costs to reward strangers who they might never meet again.

The gamut of mathematical models and experimental games by economists, psychologists, and evolutionary biologists is nicely summarized by Karl Sigmund in the provocatively titled *The Calculus of Selfishness*.[24] Even when self-interest is adopted as a guiding motivation, many social conditions predictably yield cooperation. Direct reciprocity underscores the role of repeated encounters; indirect reciprocity, the role of reputation. Incentives can foster fairness and trust. Free choice and enforcement enable joint efforts and shared public goods. Ultimately, the social trumps the individual. The higher level of organization develops its own distinctive properties. Cooperation is *emergent*. Such emergent properties sharply curtail the explanatory power of genetic reductionism, inherent in the notion of the "selfish gene."

But are such models realistic? Do humans behave according to their assumptions? A team of fourteen anthropologists investigated the willingness of individuals to punish noncooperators at a cost to themselves. Unlike earlier studies, they documented this tendency across an impressive diversity of cultures: a total of fifteen, from Pacific Islanders and African pastoralists to Siberian hunters and US college students. Furthermore, the degree of endorsed punishment correlated with what the researchers' calculated as the culture's "generosity index" (a conclusion echoed in another study, of sixteen diverse groups in six developed cultures).[25] Cooperation was correlated with readiness to punish.

A more recent, follow-up study addressed how the cultures varied in their specific conceptions of fairness. Just *how much* were others expected to share?[26] Two factors were confirmed. First, people tend to share more (that is, give closer to one-half to another person) as their culture depends more on economic markets. Apparently, societies that rely on the exchange of goods and services deem equitable transactions important and thus support a norm of fairness. Second, the willingness to punish increases with population size. As populations become larger, fleeting interactions with strangers increase. Without direct reciprocity, however, the need for trust is sharpened. For the researchers, these findings helped underscore the role of social institutions and norms—not innate tendencies based on kinship—in establishing cooperation in large societies. Once again, emergence at the social level is key. Cultures can evolve on their own and govern individual behavior.

When the Social Environment Selects

Another important thread of research has focused on moral feelings—what Darwin and his peers called the "moral sentiments." What is the origin of the impulse to help victims of a hurricane or earthquake in some remote region of the globe? Or help strangers when a skyscraper bursts into flame? Certainly not calculations of kinship. Nor anticipations of reciprocity. Psychological and evolutionary contexts) are distinct. Proximal and ultimate causal mechanisms function on different levels, and at different time scales.[27] An individual's motivational system is independent of a species-level adaptation, even if shaped by natural selection.

Empathy may possibly be learned through enculturation or explicit educa-tion—another form of social-level regulation. Yet even young children may express it spontaneously. For example, they often console others who are cry-ing. Felix Warneken and Michael Tomasello, at the Max Planck Institute for Evolutionary Anthropology in Germany, documented how eighteen-month-old infants provide help to adults when they notice simple problems as the adult tries to complete some task. Chimps do the same, although at a later age.[28] Chimps have also responded with concern to the mock crying of their human caretakers (for example, the chimp Washoe, responding to Beatrix Gardner in the 1960s; or Yoni, responding to Nadia Ladygina-Kohts in the 1930s). De Waal catalogs numerous other examples of empathy among primates, showing the ancestral roots of such responses. Chimps have ventured into water to save others (often drowning themselves). Other chimps have assisted an older colony member with arthritis. Snow monkeys tolerate a troop member born without hands. Macaques offer consolation after a sexual assault, and so on.[29] De Waal justly criticizes the "veneer theory" that humans are fundamentally selfish to the core, with only a thin surface of morality imposed by society.[30] Some empathetic motivations seem innate.

How could moral instincts evolve, if not by kin selection? Darwin provided a model of sorts in his concept of sexual selection.[31] Organisms adapt, but *to other organisms*—that is, to a *social* environment. Such selection has generated some pretty remarkable traits in reproductive contexts: colorful peacock plumage, gar-gantuan Irish elk horns, complex melodic whale songs. So, too, for ancestral human societies? Behavioral traits that enhance personal survival in a social world will (when genetically based) contribute more genes to succeeding generations. Such a process is now used to explain the origin of several human social traits: language skills, the ability to interpret each other's perspectives, and "social intelligence" (such as the ability to detect liars).[32] According to de Waal, our primate ances-tors evolved other innate social tendencies that we inherited: emotional contagion, concern for others, and conflict resolution.[33] To the degree that we are social animals, we should expect moral sentiments and cooperative tendencies to be integral to our evolved heritage. The social environment can be a selective force just as much as the physical environment.

Darwin's concept of artificial selection also seems relevant. As a gentleman farmer Darwin readily perceived how domesticated animals and their wild cousins exhibit different traits. That results from generations of intentional selective breeding. Accordingly, one might say that humans have also been "domesticated": by each other. Society will collectively "breed" for innate cooperative and social dispositions. The irony is that humans themselves create the very social environment that shapes their own social dispositions through selection.

Domesticating the Selfish Gene

In retrospect, the concept of the selfish gene, like Robert Ardrey's "territorial imperative" of the 1960s, seems like a nightmarish expression of individualism drawn from Cold War politics and capitalist economics. While it inspired much thinking (some of it fruitful), it also seemed to biologize society. It gave nature-based justification to ideological views.[34] The authority of science appeared to endorse (inappropriately) certain cultural values. "Selfishness" was *naturalized* (on naturalizing, see essays 7 and 16–20).[35] Genuine cooperation became an aberration or a lie, or at least an explanatory paradox. The recent wave of research highlighting the role of emergent properties now makes quite clear those earlier biases and distorted assumptions.

Of course, one may be equally blinkered by overly romantic views. Humans are hardly universally moral or beneficent. (Witness the atrocities at Abu Ghraib prison, enslavement of girls in Nigeria, use of chemical weapons, racial violence in the United States, or even just bullying in school settings.) Researchers have not neglected this darker side of behavior. For example, Keith Jensen has explored the conditions for spite and runaway punishment.[36] Consider also the role of oxytocin, a peptide neurotransmitter-hormone released from the hypothalamus. In recent years it has been shown to regulate mother-infant bonding in sheep and rodents, pair bonding in voles, and group size in zebra fish.[37] In humans, it seems to promote empathy, trust, and generosity while diminishing unfair exploitation. One might thus want to champion a new biological basis of—and physiological mechanism for—cooperation. Indeed, practical applications of oxytocin as a "social lubricant" are already underway, from business to personal romance to law enforcement. However, according to a Dutch study, those rosy effects do not seem divorced from antagonism toward outsiders.[38] Here, in-group loyalty seems coupled to out-group aggression. Oxytocin's regulatory role apparently has two social edges: fostering within-group helping and intergroup conflict both. Meanwhile, analysis of anthropological data has indicated that conflicts between competing hunter-gatherer groups could well have been significant in human evolution.[39]

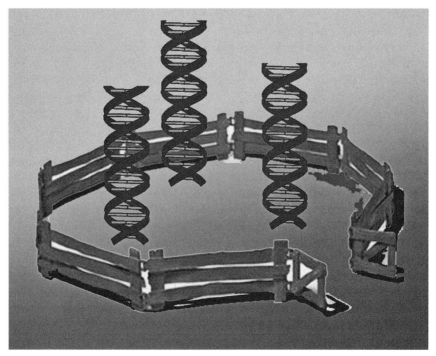

FIGURE 9.2 *The domesticated gene. Selection in a social environment can foster cooperative behavior and the evolution of moral instincts.*

Well, no surprise perhaps: biology is complex. One extreme view will not be solved by an alternative extreme.

As research continues, knowledge grows, yes. But sometimes, concepts can change dramatically. The overall gestalt can switch. Paradigms can shift.[40] Sacred bovines can topple. Genocentric views of cooperation now seem quite limited. Instead, genes, minds, and society interact, integrating different levels. In particular, higher levels of organization can govern lower levels. The once-faddish "selfish gene" has become domesticated (Figure 9.2).

PART III

Making Mistakes

10

A Comedy of Scientific Errors

William Shakespeare may well have foreshadowed the modern television sitcom. His comic misadventures were expertly crafted. In *A Comedy of Errors,* for example, twins with twin servants, each pair separated at birth, converge unbeknownst to each other in the same town. Mistaken identity leads to miscommunication. More mistaken identity follows, with more misdelivered messages and yet more misinterpretations. Hilarious consequences ensue. It is a stock comedic formula in modern entertainment: A character first makes an unintentional error. Then ironically, as he tries to correct it, things only get laughably worse.

Science, we imagine, is safeguarded against such embarrassing episodes. In the lore of scientists, echoed among teachers, science is "self-correcting." Replication, in particular, ensures that errors are exposed for what they are. Research promptly returns to its fruitful trajectory. Serious stuff, science.

But just such a case of compounded error occurred in late eighteenth-century science. Joseph Priestley (Figure 10.1) discovered that plants can restore the "goodness" of air that had been fouled by animals or combustion. But others could not replicate his results. Not even Priestley himself. After further work, Priestley attributed the observed restorative effect to a different causal factor—only to find later that the new conclusion itself was mistaken! For us now, the story seems amusing, but nonetheless instructive. The case invites us to reconsider the sacred bovine that science is self-correcting, and especially that replication is central to exposing errors. Indeed, this reassessment leads us deeper into reflecting on our romantic idealizations of science, an enduring legacy of Priestley's Enlightenment period, centuries ago.

* * *

The story begins in the early 1770s, in Leeds, England.[1] Joseph Priestley—minister, avid experimentalist, and self-taught chemist—had been investigating various kinds of air. At this time, he was examining various ways of making air noxious: by the putrefaction of dead mice or cabbage, by burning charcoal in it, by mice breathing it, or by candles burning out in it (all processes that exhaust the oxygen, in today's terms). Such modified air would not support animal respiration. What was

FIGURE 10.1 *Joseph Priestley, whose mistaken interpretations on the restoration of air by plants led to a comical cascade of errors.*

the nature of this air, and how might its deficit be remedied? Priestley, who liked to tinker with variations of his experiments, investigated the possible airs emitted by plants as well. He later recalled: "On the 17th of August 1771, I put a sprig of mint into a quantity of air, in which a wax candle had burned out, and found that, on the 27th of the same month, another candle burned perfectly well in it. This experiment I repeated, with the least variation in the event, not less than eight or ten times in the remainder of the summer."[2] Then he tested just oil of mint, to see if the effect was caused merely by the plant's aromatic "effluvia." It was not. Subsequently, he tried the experiment with other plants—balm, groundsel, and spinach. All modified the air to support sustained burning. Animals, too, could breathe longer in the treated air.

Plants, Priestley had found, could restore the "goodness" of the air depleted by respiration or combustion. He wrote to Benjamin Franklin in America, who immediately perceived the global implications: plants help restore the atmosphere that humans and other animals foul. The system ensures our survival. That view fit neatly with Priestley's religious belief in an intentionally designed (and rational) natural world. It was a remarkable discovery. For this and other work on airs, the Royal Society in 1772 awarded Priestley the Copley Medal, then the most prestigious honor in science.

Others were eager to build on Priestley's discovery about plants and the restoration of air. But they could not always get the same results. Today, we might say that they frequently failed to replicate his experiment. That led to some confusion. Priestley returned to his experiments himself, but only a half decade later. By then he had moved to a new city—and a new experimental workspace. Like others, he could not consistently obtain his earlier results. Indeed, in some cases, the plants now seemed to *worsen* the quality of the air! His original claims seemed awkwardly in question. Should he retract them? Priestley had already received the Copley Medal and his findings had been praised by the president of the Royal Society. And the original conclusions fit comfortably with his worldview. He thus discounted the significance of the negative results: "one clear instance of the melioration of air in these circumstances should weigh against a hundred cases in which the air is made worse by it."[3] Once the "discovery" had been made, Priestley seemed reluctant to acknowledge it as an error.

Priestley persisted. Following his habits, he explored more minor experimental variations, without any major theory or hypothesis to guide him. Eventually he noticed an apparent role for sunlight. In his first set of studies, he had apparently missed the role of a nearby window, absent in his new workspace. Priestley now had a new relevant variable: light. At this point in the story, the informed modern observer may anticipate next the triumphant discovery of the role of light in photosynthesis. With the error virtually solved, one is poised for a "happily ever after." But here the comedy of errors unfolds differently. Perhaps, Priestley wondered, light alone—*not* plant life—was key? Accordingly, he tried simple samples of well water exposed to sunlight, without plants in them. They, too, yielded bubbles of the "purer," more respirable air. Priestley now felt confident that he had identified the source of error in his original work. The process of restoring the air, he concluded, was related to light, not plants! Error resolved. Or so it seemed, ironically, to Priestley.

Of course (from a modern perspective), it was the newly revised conclusion that was in error. Here, the scene shifts to others working on the problem. Jan Ingenhousz and others noticed that when well water was left in sunlight, it not only produced bubbles of restored air, it also generated a *green scum*. They connected the green scum to green plants. With further microscopic analysis they concluded that the scum was living algae. So plants or microscopic algae—both green living matter—had transformed the air. But only in light, they now realized. Ingenhousz demonstrated the connection more fully through an extensive series of controlled tests. Both green plants and light together were needed to restore the air, not one or the other. Ingenhousz, and then others, also perceived that the plants producing good air in light was opposite to burning them, using up good air, and releasing light. The plants were absorbing the light somehow to make fuel. Moreover, the process coincided with restoring the air. It was the reverse of combustion. Here, emerging in part from Priestley's successive errors, was the discovery of what we now call photosynthesis.

Priestley had noticed the green scum too. But he had considered it a by-product of the enriched atmosphere. No light, no bubbles; no bubbles, no scum. In retrospect,

Priestley's experimental results were ripe for mistaken identity. Correlation could resemble causation, in two ways (Figure 10.2). First, the light seemed directly responsible for the restored air. Priestley saw, but discounted the significance of, the correlated green matter. Second, the enriched air seemed to cause the green scum, not the other way around. If we laugh, it is because we can see how easily we, too, could have been mistaken. To his credit, Priestley acknowledged his error, once the new explanation had been clearly demonstrated. Error remedied, lessons learned, plot resolved. Scene fades as the wistfulness of comedic humility lingers.

* * *

Priestley's successive misinterpretations offer an opportunity to reflect on error in science. According to standard accounts, replication is the chief method for identifying experimental error. Failure to replicate means one should jettison the results as wrong. In this case, that would have been ill advised. Priestley's original findings were correct. Rather, the "failed" replications were problematic. Ultimately, Priestley did not know at first—and was thus unable to specify to others—the exact conditions under which his plants had restored the air. When the role of light became clear, replications "succeeded." But even then, repetition alone did not confirm Priestley's new conclusion about the exclusive role of light was correct.

Priestley's errors did not merely announce themselves. Contrary to popular expression, the data do not speak for themselves. Observations need to be interpreted. Errors, too. Here, finding and characterizing the error required further scientific work. Priestley and others had to identify both light and green plants as key factors. Ingenhousz's controlled studies were needed to isolate the relevant variables and to demonstrate their significance. In a sense, he had to replicate Priestley's *errors* successfully while *also* showing what caused the errors. Fixing errors in science is

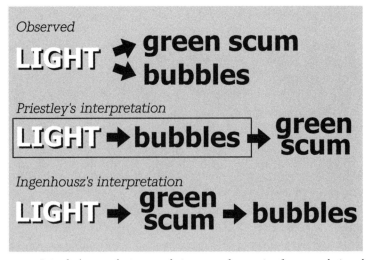

FIGURE 10.2 *Priestley's error: he incorrectly interpreted causation from correlation alone.*

not just about discrediting or discarding "negative" results. Paradoxically, perhaps, it involves understanding them. At the same time, this yields new knowledge.

One may reflect further on just how Priestley's error was discovered and remedied. First, the process required motivation and resources. Priestley had wealthy subscribers and the patience to persist at endless variations of his experiments. Ingenhousz, too, had both the interest and leisure time to devote to research. Science proceeds concretely and materially, not through imagination or ideas alone. Second, identifying the green scum as plant life required a microscope. That was a technological contribution. One also needed the disposition and skills to use the microscope, which Priestley largely lacked but others supplied. Finally, interpreting the correlation of light and green scum depended on a repertoire of alternative theoretical perspectives, here shared across the scientific community. The case of Priestley's errors and their resolution ultimately helps convey the nature of science, or how science works.

According to standard accounts, science is self-correcting. While Priestley's errors were indeed ultimately corrected, no systematic or automatic "self-correcting" process was responsible. The revision took several years of focused effort. Error correction cannot be taken for granted.

The image of science as self-correcting reflects a kind of rosy intellectual optimism that flourished among the elite in Priestley's era and has largely persisted since. Benjamin Franklin expressed the view well: "Truth is uniform and narrow; it constantly exists, and does not seem to require so much an active energy, as a passive aptitude of soul in order to encounter it."[4] For Franklin (and many others), one need not explain the emergence of knowledge. Insight is supposed to unfold naturally. And effortlessly. Priestley's case shows how this naive Enlightenment view, still embraced by some science educators, is ill informed.[5] Sources of error and occasions for "mistaken identity" permeate science. It requires work to articulate what's right and to sort out what is the case from what merely appears to be so. That is how we generate knowledge. Resolving Priestley's errors was thus largely also the story of discovering photosynthesis.

Priestley's mishaps might amuse us today, but the reflective practitioner laughs with him, not at him. His mistakes remind us of the very human dimension of science—and of our own potential for error.

11

Nobel Ideals and Noble Errors

Christiaan Eijkman shared a 1929 Nobel Prize "for his discovery of the antineuritic vitamin." His extensive studies on chickens and prison inmates on the island of Java in the 1890s helped establish a white rice diet as a cause of beriberi, and the rice coating as a remedy. Eijkman reported that he had traced a bacterial disease, its toxin, and its antitoxin. Beriberi, however, is a nutrient deficiency. Eijkman was wrong. Ironically, Eijkman even rejected the current explanation when it was first introduced in 1910.[1] Although he earned a Nobel Prize for his important contribution on the role of diet, Eijkman's original conclusion about the bacterium was just plain mistaken.

Eijkman's error may seem amusing, puzzling, or even downright disturbing— an exception to conventional expectations. Isn't the scientific method, properly applied, supposed to protect science from error? And who can better exemplify science than Nobel Prize winners? If not, how can we trust science? And who else is to serve as role models for students and aspiring scientists?

Eijkman's case, however, is not unusual. Nobel Prize–winning scientists have frequently erred.[2] Here I profile a handful of such cases (Figure 11.1). Among them is one striking pair, Peter Mitchell and Paul Boyer, who advocated alternative theories of energetics in the cell. Each used his perspective to understand and correct an error of the other! Ultimately, all these cases offer an occasion to reconsider another sacred bovine—that science is (or should be) free of error, and that the measure of a good scientist is how closely he or she meets that ideal.

An Error for Every Nobel?

Consider first Linus Pauling, the master protein chemist.[3] Applying his intimate knowledge of bond angles, he deciphered the alpha-helix structure of proteins in 1950, which earned him a Nobel Prize in 1954. He also reasoned fruitfully about sickle cell hemoglobin, leading to molecular understanding of its altered protein structure. Yet Pauling also believed that megadoses of vitamin C could cure the common cold.[4] Evidence continues to indicate otherwise, although Pauling's legacy still seems to shape popular beliefs.[5] His unqualified advocacy eventually

FIGURE 11.1 *Ten Nobel laureates, who all erred: Christiaan Eijkman (a), Linus Pauling (b), Albert Szent-Györgyi (c), Paul Ehrlich (d), Ilya Mechnikov (e), James Watson (f), Francis Crick (g), John Eccles (h), Peter Mitchell (i), and Paul Boyer (j). To err is science?*

led to his losing sources of financial support. Pauling sometimes described the source of good ideas as having lots of ideas and throwing away the bad ones. That may well characterize science. Yet it highlights the question of how one recognizes bad ideas and how long they may linger, with what effect, before being thrown away.

Pauling's ideas about vitamin C partly echoed those of another Nobel Prize winner, whom he called "the most charming scientist in the world": Albert Szent-Györgyi.[6] Szent-Györgyi isolated vitamin C and helped identify it as ascorbic acid. Later, he buoyed research by showing how vast quantities of it could be extracted cheaply from the paprika peppers of his native Hungary. He also 'claimed, erroneously, that vitamin C participates as an intermediate in mitochondrial reactions in the cell and that it could cure various medical conditions. Szent-Györgyi received a Nobel in 1937 "for his discoveries concerning the biological combustion processes." He had helped resolve a debate about those reactions—showing how oxidations leading to proton transfers could be reconciled with electron flow and the use of oxygen. He also helped elucidate the cellular role of fumaric acid (although he identified it incorrectly as a catalyst, rather than a chemical intermediate, in a reaction series). Szent-Györgyi went on to contribute to muscle physiology, demonstrating the role of energy in the interaction of actin and myosin molecules. Yet he also made many spurious claims, such as having discovered yet another vitamin (vitamin P). He promoted treating diabetes with succinic acid and cancer with ultrasound or mushroom juice! For every fruitful idea Szent-Györgyi offered, it seems, there was at least another that was equally mistaken. Given his heroic renown, of course, the errors often remain in shadow.

The 1908 Nobel Prize in Physiology or Medicine marked a pair of discoveries— and perhaps a pair of errors. Paul Ehrlich had characterized the immune reactions of agglutination, bacteriolysis (via complement), and hemolysis. His work embodied the then-popular approach to immunity, which focused on blood chemistry. (Only seven years earlier, Emil von Behring had received one of the first Nobels "for his work on serum therapy"—with its hope for curing all infectious disease through transfusions of blood sera.) Ehrlich, however, also denigrated cell-oriented approaches as utterly misguided. He erroneously excluded any role for white blood cells engulfing waste, say, or for immune action mediated by what we now know as T-cells. Such processes had already been observed and investigated by Ilya (Élie) Metchnikov—who shared the 1908 Nobel with Ehrlich. Metchnikov, in his turn, erred in dismissing the promise of approaches based on chemical elements carried in the blood. The prize committee recognized the complementary contributions together, signaling that it regarded the respective claims of limited scope as unfounded.[7]

What of the famed cocreators of the double helix model of DNA, James Watson and Francis Crick? Surely *they* did not err? Having established the structure of DNA in 1953, they went on to probe the relationship between DNA and proteins and to interpret its "genetic code." In 1958 Crick proposed a theoretical guidepost: "Once information had passed into protein it cannot get out again."[8] This "central dogma" became expressed in Watson's 1965 book, *Molecular Biology of the Gene*, as

$$DNA \rightarrow RNA \rightarrow protein$$

Watson's simple formula gradually eclipsed Crick's and gained widespread currency as expressing a family of truths beyond doubt. First, the cellular functions of information (inheritance) and enzymatic catalysis (metabolism) were differentiated into distinct molecular types. Second, only DNA could self-replicate. Third, information flowed irreversibly from DNA nucleotide sequences through RNA to amino acid sequences. All three principles later yielded to exceptions—although not without controversy. Indeed, it is a measure of the depth of this ensemble of errors that each counterdiscovery itself earned recognition. The Nobel Foundation honored Howard Temin and David Baltimore in 1975 for discovering reverse transcriptase—which produces DNA from RNA; Sidney Altman and Thomas Cech in 1989 for discovering ribozymes—RNA that can fold on itself and catalyze certain reactions; and Stanley Prusiner in 1997 for characterizing prions—proteins that can "reproduce," or at least provide the "information" to transform similar proteins into new, disease-causing agents. (The 2006 prize announcement for Andrew Fire and Craig Mello implied that RNA interference, too, violated the central dogma—by interrupting the "normal" transfer of information from RNA to protein.[9]) All these discoveries indicated that the "dogma" of the "central dogma" was ill conceived.

Francis Crick, for his part, never advocated all the wrong ideas implied by Watson's expression. Crick himself was not completely free of error, however. He seems to have not understood fully the meaning of the word "dogma." He chose it when he meant something very different: an unjustified belief, rather than an inviolable tenet.[10] Crick tried to clarify his meaning in 1970, but the very label continued to signify to others that they should regard the central dogma erroneously as—well, dogma.[11] Crick earns note for yet another, more substantive error. He became increasingly impressed by the complexity of the cell's protein-making process. He could not imagine the circumstance under which it could have originated here on Earth. Thus, in 1981 he endorsed the notion of panspermia—that life originated elsewhere and arrived here by deliberate (though unspecified) means.[12] Quite understandably, scientists did not receive this maverick idea with the same esteem and respect as the double helix model.

Finally, consider John Eccles, recognized in 1963 for discoveries related to the "ionic mechanisms ... of the nerve cell membrane." More specifically, Eccles helped characterize the transmission between neurons. At some cell-cell junctions, he found, an impulse hyperstimulates the receiving cell's membrane, thereby inhibiting successive impulses. At other junctions, by contrast, the impulse enhances the membrane's receptivity. When the sensitivity is lowered sufficiently (to some threshold) from multiple impulses or junctions, the next neuron starts its own impulse. That also indicated that the junctions between nerve cells function chemically. Since nerves can only fire or not fire, the chemical mechanism for combining positive and negative signals is critical to producing complex responses. Eccles thereby helped elucidate the biological basis of mind.[13] Yet Eccles also wrote extensively that mind and body were distinct. Ultimately, he argued that the existence of a divinely created soul was grounded in science. While one might localize the phenomenon of consciousness, he contended, there were still connections with another (nonmaterial) entity to be described. For Eccles, biology could not explain free will. He could not reconcile strict determinism with the concept of moral responsibility.[14] Eccles applied his dualist view to the evolution of the brain, asserting that "there can be no physicalist explanation of this mysterious emergence of consciousness and self-consciousness in a hitherto mindless world."[15] Today, one can only wonder at how Eccles tried to deploy naturalistic science to nonnaturalistic ends, ironically in his own field of expertise, neurophysiology.

Exchanging Errors?

Scientists, it seems, do not always recognize their own errors. That task seems to fall to other scientists. One can thus imagine a circumstance where two scientists could possibly "return the favor" by correcting each other's mistakes. One such case seems to have occurred in cellular bioenergetics over several decades late last century.[16]

The first error was made by Paul Boyer in 1963. (Fret not! He would earn a Nobel 34 years later.) In the 1950s biochemists were looking for a set of high-energy molecules that transferred energy from the electron transport chain to the final energy molecule (adenosine triphosphate, or ATP). After a decade of failed claims from several labs, Boyer reported evidence in the prestigious journal *Science* that he had isolated the intermediate and identified it as phosphohistidine. Relief cascaded through the community. The high-profile triumph was short-lived, however. Boyer's lab soon attributed the results to other energy reactions in the cell. (The data were "real," but when contextualized with other results, dramatically reinterpreted.) "I was wrong," Boyer later put it bluntly.

Boyer was actually wrong on two levels at once. Phosphohistidine was not the intermediate. Boyer admitted as much. But the very concept of the intermediates, for which everyone had been searching so earnestly, was also mistaken. Boyer soon reached that conclusion as well. (If he hadn't found the intermediate using his methods, he boldly believed, no one else would.) Boyer hypothesized instead that the energy must be transferred through energized changes in protein shape (like a pair of interacting molecular springs). That concept, too, would eventually prove mistaken.

Here, the unexpected solution was introduced by Peter Mitchell. Mitchell was guided in his thinking by a novel principle of vectorial chemistry—that enzymatic reactions happened spatially (for example, with reactants and products on different sides of a membrane). Synthesizing many clues, Mitchell conceptualized the intermediate energy state as a proton gradient across a membrane—a chemiosmotic potential. That revolutionary idea ultimately earned Mitchell a Nobel Prize in 1978.

Mitchell's own claims, however, were hardly free from error. In the first formulation of the theory, for example, the direction of the chemiosmotic gradient was reversed! Mitchell also specified one proton everywhere two were needed. Such "minor" errors were soon remedied. But the unrealistic quantitative analysis had already convinced many chemists that Mitchell's notions were fundamentally flawed.

Most dramatically, Mitchell had a vision about how ATP was synthesized from the proton gradient. Using his foundational principle of vectorial chemistry, he insisted that protons flowed into the interior of the ATP enzyme and there participated directly in forming the energized phosphate bond of ATP. That creative concept never fit comfortably with the data. Here, it was Boyer's concept, rather, that prevailed. Boyer had adapted his ideas on the role of protein shape. He reasoned how ATP formed on the surface of the enzyme and was then released through an energy-requiring change in the enzyme's shape. The energy was provided remotely by protons rotating the enzyme as they recrossed the membrane to lower energy levels. Those insights were recognized in a 1997 Nobel Prize. Ultimately, Boyer and Mitchell had both been right (partly). And both had been wrong (partly). Their perspectives neatly complemented each other's blind spots.

Role Models and Real Models

Well, what is one to make of all this error among the world's most highly regarded scientists? Do these examples make science entirely meaningless? Hardly. We cannot discount the great discoveries. Nor their great discoverers. Indeed, the errors seem informative just because the scientific credentials of those who made them are unassailable.

Ultimately, if Nobel Prize winners can be mistaken, then any scientist can be. Science is a human endeavor. Of course, no human is perfect. "To err is science," we might say.

Still, healthy science can root out error. As the cases of Boyer and Mitchell (or Ehrlich and Metchnikov) exemplify, contrasting views cross-check each other. They promote completeness of evidence. The chief safeguard against persisting error, then, seems to be not blind skepticism but actively engaged diversity. Science is empirical, but it is also, ideally, social.[17]

What does all this portend for inspiring the next generation of scientists? How can we motivate careers in science if we cannot celebrate Nobel Prize winners as role models? Historian Stephen Brush once wondered if less-than-ideal portrayals of scientists shouldn't be restricted to "mature audiences" only. "Should the history of science be rated 'X'?" he asked (alluding to the then-new film rating system).[18] Brush was ambivalent. The heroic image, he suggested, might contribute to recruiting future scientists. On the other hand, the human dimension also seems valuable, for educating nonscientists. Awareness of flaws fosters a deeper understanding of the nature of science. The dilemma has renewed vigor with policies now aimed to promote more careers in science. Do we portray real scientists, mistakes and all, or more-inspirational but fictional ideals?

Ultimately, such a choice reflects a false assumption. Why suppose that role models should be flawless? Why expect that great individuals, like those profiled above, never make mistakes? Why is making errors not noble? The errors that should concern us are not those of the scientists themselves but our wildly idealized yet widely held expectations of them.

Role models need to be realistic. Indeed, human-scale role models—"real models"—may well bring more esteem to science and generate less disillusionment than scientific fairy tales. We owe our maturing youth plentiful inspiration, sustained encouragement, and well-informed guidance, not phantom goals.

An understanding of science is incomplete without acknowledging that scientists—even Nobel Prize winners—can err. We may equally want to highlight that such errors are generally found and remedied through the social structure of science. That lesson may underscore the value of another principle: not just nurturing prospective Nobel laureates but also building a diverse, balanced scientific community.

12

Celebrating Darwin's Errors

Charles Darwin was truly amazing. In 1859 he introduced a robust understanding of descent with modification by means of natural selection. His concepts would help unify taxonomy, biogeography, comparative anatomy, heredity, functional analysis of form, embryology, paleontology, population dynamics, and ecology, and even human moral behavior. Darwin showed how to explain organic "design" as well as the limitations of contingent history, adaptive structures as well as vestigial ones. Every lesson in biology, properly framed, expresses and celebrates Darwin's achievement.

How, then, might one mark so august an occasion as his two hundredth birthday (also the sesquicentennial year of his premier work, the *Origin of Species*)?[*] Many will no doubt parade Darwin's many triumphs. But allow me to take exception to the common view (another sacred bovine?) that science is best reflected only by its successful theories. If science is fundamentally about discovery, then its "failures" or errors along the way may be just as important as the ultimately reliable insights. I wish to celebrate science *as a process*. Here, then, I acknowledge Darwin's mistakes and show how understanding them gives us a deeper understanding both of Darwin and of science more generally. My tribute is to forgo the mythologized legend and appreciate so remarkable a scientist as Darwin in familiarly human terms.

Was Darwin Ever Wrong?

First, one may note that Darwin's errors generate interest largely because of his many achievements. His credentials are unimpeachable. If he made mistakes, it was not for want of scientific ability. One cannot rudely dismiss his errors as due to ineptitude.

Indeed, Darwin's contributions are wider and their theoretical coherence deeper than popularly known.[1] He produced four volumes on the taxonomy of barnacles,

---

[*] This essay was originally written in 2009 to help mark Darwin's bicentennial.

demonstrating his skills in detailed observation and analysis of evolutionary classification. In his first work after the *Origin*, he showed the importance of orchid form in promoting outcrossing through pollination, thereby contributing to an understanding of the role of sex and genetic recombination in evolution. Later, he explained heterostyly—the occurrence of flowers with styles of different lengths—as illustrating the same general principle (see essay 16). Add, too, his work on the anatomy and physiology by which emotions are expressed, grounding a study of mental phenomena and social communication in concrete observables. In his last work, Darwin correctly interpreted the role of worms in forming topsoil (what he called "vegetable mould").

Darwin was also a skilled experimentalist.[2] Chapter 11 in the *Origin* summarizes some of his experiments on the effects of seawater on seed germination—a small test of his ideas about how plants could travel across an ocean. With his son Francis he investigated "the power of movement in plants"—documenting and measuring phototropism, and determining how it occurred in certain parts of the plant. These studies followed earlier experiments on the positive effects of plant hybridization. Darwin would surely be remembered for these works even if he had never written the *Origin* or *The Descent of Man*.

In short, there is no deficit of Darwin's achievements.

Yet Darwin's conclusions were not always correct. Perhaps the most notorious of his ill-fated claims was his "retreat" to Lamarckian-like influences.[3] While variation was essential to the process of natural selection, Darwin could not explain its sources. Sharp criticism worsened the problem. Darwin, rather than perhaps leave his theory incomplete, ultimately appealed to external forces (use or disuse, or habit, say) in generating favorable variants. That seemed to echo Lamarck's earlier idea (now discredited) that acquired characters could be inherited. Darwin also claimed that domestication of wild animals itself increased the rate of generating new variants.

Many admirers today wonder: How could the great Darwin have succumbed to such nonsense? Indeed, modern portrayals of Darwin often treat this politely as a blemish or mild embarrassment. They tend to excuse it as a product of the times. (What idea is not a product of its time?) Or they downplay Darwin's level of commitment, implying that he didn't *really* believe in it. Of course, such dismissals never extend to Darwin's correct claims. Historical judgments are easily shaded by later outcomes. Too often, we tend to manipulate the past to fit our own ideals. We render the science as more perfect than it really was—or is now.

But Darwin professed what he professed. Other options were available at the time. Indeed, the codiscoverer of natural selection, Alfred Russel Wallace, saw no need to *explain* variation. He chided Darwin in a letter in 1866:

Such expressions have given your opponents the advantage of assuming that *favorable* variations are *rare accidents*, or may even for long periods never occur at all and thus [the] argument would appear to many to have great force. I think it would be better to do away with all such qualifying expressions, and constantly maintain (what I certainly believe to be the fact) that *variations of every kind are always occurring in every part of every species,* and therefore that favorable variations are *always ready* when wanted.

For Wallace, the mere fact of variation was enough to answer critics. He continued:

> You have, I am sure, abundant materials to prove this, and it is, I believe, the grand fact that renders modification and adaptation to conditions almost always possible. I would put the burden of proof on my opponents to show that any one organ, structure or faculty, does *not vary,* even during one generation, among the individuals of a species and also to show any *mode* or *way,* in which any such organ, etc. does not vary.[4]

Wallace is a convenient touchstone for assessing Darwin's error on this occasion.

Darwin made other mistakes, as well—some trivial, some less so. First (ironically), Darwin failed to properly label the finch specimens he collected on the Galápagos Islands: the very species that would later bear his name. Ornithologist John Gould, who worked on his collection, noticed the error and helped remedy it by consulting further specimens collected by others on the *Beagle* voyage.[5]

Later, having established descent with modification as a general doctrine, Darwin endeavored to fill in some of the details. Here, his proposals met with mixed success. Darwin proposed that modern chickens are descended from red-footed junglefowl. Recently, geneticists have identified the foot color gene. They have determined that chickens get their yellow feet, ironically, from having hybridized with *gray* junglefowl.[6] (Religious critics of Darwinism had a field day with this little blooper!) Darwin erred, too, in thinking of the fossil *Eozoon* as primitive biota that helped fill the apparent gaps in the early history of life. Further analysis revealed it to be an inorganic mineral formation, as Darwin himself acknowledged.[7] These errors are all relatively minor. Yet they remind us that small mistakes occur commonly in science. When findings become important enough, follow-up studies tend to either confirm earlier results or reveal how perceived patterns were based on incomplete information.

Biases in Discovery

Darwin's errors (like those of other great scientists) can often be coupled to one of his notable discoveries. The paired conclusions ironically drew on the same underlying concept or exhibited the same style of thinking. Each case highlights Darwin's distinctive perspective (or "bias," perhaps). Sometimes, then, erroneous ideas and successful ideas had a common origin.

Consider two of Darwin's early theories in geology. Both applied Charles Lyell's principle of uniformitarianism—viewing the past as a cumulative product of gradual forces still present today. In the first case, Darwin addressed the natural history of coral atolls. He reasoned that reefs formed around islands, which then gradually eroded, leaving hollow rings. It was an act of sweeping historical imagination based on observational fragments about coral growth and location. The idea helped launch Darwin's career—and it proved correct.[8]

FIGURE 12.1 *The parallel roads of Glen Roy. Darwin concluded, incorrectly, that these geological formations were the shores of a receding ocean. However, Darwin was relying on the same style of reasoning that allowed him to successfully interpret the geological origin of coral atolls. Darwin later acknowledged that these rock shelves were remnants of retreating glaciers.*

Darwin applied the same kind of large-scale gradualist thinking to the "parallel roads of Glen Roy," a series of stony ledges lining a valley in Scotland (Figure 12.1). He imagined that they were the debris of successively lower shorelines, left by a receding ocean. Here, Darwin was wrong. The ledges were moraines, left by a retreating glacier, not an ocean. Darwin, to his credit, acknowledged his "great blunder" when Louis Agassiz's theory of glaciation and ice ages gained prominence.[9] Darwin was right and wrong, on different occasions, by relying on the same Lyellian reasoning in both cases.

A second major discovery intimately combined with error concerns Darwin's reasoning about human descent. Darwin's gradualism fostered much productive thinking about transitional forms—for example, in his evolutionary classification of barnacle sexual systems. Yet the concept had especially powerful implications in the context of his social status. British society was stratified. Darwin enjoyed membership in the upper class. He was also a white European at a time when Europeans (notably the British) dominated the globe. This context shaped perceptions of other races, easily construed in a hierarchy. While voyaging on the *Beagle*, for example, Darwin was appalled by the habits of the natives of Tierra del Fuego (Figure 12.2): "It was without exception the most curious and interesting spectacle I ever beheld: I could not have believed how wide was the difference between savage and civilized man: it is greater than between a wild and domesticated animal, inasmuch as in man there is a greater power of improvement."[10] Improvement there was. One of the Fuegians had been taken to London, educated, and entered into elite society. When he returned, however, he seemed content to revert (as Darwin

FIGURE 12.2 *A native of Tierra del Fuego. Darwin's assumptions about progress and races led him to view the aborigines incorrectly as an evolutionary transition between primates and more-civilized humans, exemplified by his elite British peers.*

saw it) to his "primitive" habits. It was all too easy for Darwin to consider racial differences as inherent and to rank them on a scale from "savage" to "civilized." That conception proved both fruitful and dramatically misleading.

When Darwin began considering human ancestry, he saw immediately that the problem was not primarily anatomical. Humans had long been classified as primates. The challenge was accounting for the origin of mental faculties and moral sensibilities. Darwin's early musings turned to the Fuegian episode. He wrote to himself in the fall of 1838:

> Nearly all will exclaim, your arguments are good but look at the immense difference between man,—forget the use of language, & judge only by what you see. compare, the Fuegian & Ourang & outang, & dare to say difference so great . . . "Ay Sir there is much in analogy, we never find out." (*M Notebook*, p. 153)

Darwin essentially cast the Fuegians as intermediates between orangutans and "fully developed" humans, such as himself and his peers. Darwin's ability to stratify races facilitated his linking apes and humans through a series of gradual changes. "Savages" became convenient transitional forms in moral and mental development.[11]

The relevance of Darwin's social status and experience sharpens when we compare him, once again, with Alfred Wallace. Wallace came from the working class. While collecting in the Malay Archipelago, he learned to respect the natives' local knowledge and benefitted from their assistance. In 1855 he wrote to a friend, "The more I see of uncivilized people, the better I think of human nature and the essential differences between civilized and savage men seem to disappear." If even such "brutes" could show kindness, Wallace reasoned, then all humans apparently shared an undiluted moral sense. He echoed those sentiments in 1873: "We find many broad statements as to the low state of morality and of intellect in all prehistoric men, which facts hardly warrant." Wallace, in contrast to Darwin, saw moral and mental discontinuity between man and beast. Wallace certainly acknowledged that humans had primate ancestry—anatomically. Still, he maintained that the human mind was unique and emerged by some guided process other than natural selection. Wallace never considered, as Darwin did, the evolution of morality (essay 6). Wallace erred in that. At the same time, however, he did not succumb to Darwin's error—viewing races hierarchically.[12]

Darwin's insight—the evolution of cognitive abilities and the moral sense—was thus partly due to an error, ranking races biologically. We now explain the origins of human culture and ethics with quite different benchmarks.[13] Appreciating the origin of Darwin's error is significant for a complete understanding of science. It should not surprise us, perhaps, that Darwin's view of human origins was taken by others to support racial ideologies, however inappropriately so.[14] The seed for that view was in Darwin's own thinking. Darwin was not politically racist, however. He and his whole extended family denounced slavery, for instance.[15] Biological facts (erroneous or not) do not themselves justify value judgments. But that does not prevent people from committing fallacies in reasoning. Darwin's error had major cultural consequences, although not by Darwin's own hand.

A third major discovery-mistake pair stemmed from Darwin's views on competition.[16] Those views also had cultural roots. Victorian England exhibited widespread poverty and great disparities in wealth. The social inequity was considered (by the wealthy, at least) as a "natural" outcome of competition for resources. Thomas Malthus in his 1801 *Essay on Population* portrayed food as inevitably limited and competition unavoidable. Reading that essay in 1838 prompted Darwin's insight on natural selection. Darwin transformed the cultural notion of a "struggle for existence" into a creative organic force. For him, competition fueled the logic of differential survival and adaptation.

But Darwin overstated the role of competition. He also saw it causing the origin of species. Competition within a species, he imagined, would promote specialization. With continued competition, specialized forms from the same population would ultimately diverge. Darwin seemed deeply impressed by the power of competition: "One may say there is a force like a hundred thousand wedges trying [to] force every kind of adapted structure into the gaps in the economy of nature, or rather forming gaps by thrusting out weaker ones" (*D Notebook,* pp. 134e–135e).[17]

Modern evolutionary biologist Ernst Mayr, however, has faulted Darwin for advocating what he calls an undemonstrated form of sympatric speciation.[18]

Similarly, geochemist Kenneth Hsü claims Darwin inaccurately portrayed competition between species as the chief cause of extinction, thereby obscuring geophysical events (especially relevant in mass extinctions).[19]

While competition may surely lead to selection, not all selection need be based on competition. Darwin framed even differential reproduction (sexual selection) as competition: competition for mates. Yet radiation of forms in new adaptive zones, for example—so nicely exemplified by the Galápagos finches—results more from opportunity in new niches than from competitive elimination. Other times, species seem to sustain themselves merely by holding on in extreme environments. Nor did Darwin seem oriented to appreciate the importance of random, indeterminate processes. Viewing life competitively both enabled Darwin to discover natural selection—and blinkered him from seeing its limits clearly.

Understanding Darwin's erroneous thinking about competition is important for our culture today. For many, Darwinism implies what is inaptly called "Social Darwinism"—namely, Herbert Spencer's social ideology of unrestrained competition (essay 7). In particular, the metaphor of Spencer's phrase "survival of the fittest" haunts our culture, even among individuals who have not learned biology (essay 8). Again, roots of such misapplied concepts may be found in Darwin and his use of language. Noting the source and scope of the error can help us establish a more informed conception of natural selection and distinguish it from cultural ideology.

Learning from Error

So Darwin could be wrong. Even about some important things. So what? The errors certainly do not justify rejecting all Darwin ever wrote. Nor should they tarnish his image. Rather, I think, they give Darwin a fine, human patina.

Understood together, Darwin's many mistakes also offer valuable lessons about the nature of science. First, what leads to error in science? As illustrated in the cases above, sometimes the very same thing that leads to discovery! Darwin's unique viewpoint was critical to both his insights and his blind spots. Fresh perspectives always have potential. Yet success is not guaranteed. We might thus be wary of some mystical property called "genius" that purports to yield unqualified insight. Darwin's discoveries and errors came from identifiable life experiences. Science is thus likely to benefit from diversity of backgrounds. Still, generating new ideas, while essential, is only half of science. The ideas—some right, some wrong—must also fit with a typically growing reservoir of relevant information.

Second, how is error in science remedied? Deeper evidence, of course. But that truism does not tell us, more importantly, how the new evidence is found. Alternative perspectives were needed to cross-check Darwin's original ideas. Agassiz's experience in the Swiss Alps was an important complement for those who had encountered glaciers remotely, if at all. Wallace's background in the lower classes was integral to counterbalancing assumptions about social hierarchy. Friedrich Engels, likewise, from his new communist perspective, was well situated to see economic ideology reflected in Darwin's theory (essay 7). To function effectively, science needs alternative perspectives—from various cultures, social classes, genders, disciplines, biographical backgrounds, and so on. Contrasting views help highlight deficits in the evidence or expose conceptual blinkers. Once again, we should not fail to notice the collective, social dimension of science (and with it, the value of diversity among scientists).

Finally, if we understand how errors occur, can't we eliminate them from science? Isn't the whole point of science to escape error and provide trustworthy knowledge? Here, one virtue of studying history may emerge. Science itself seems structured like Darwin's concept of natural selection. It balances novel conceptual variants with selective retention via testing and other forms of checks and balances. No algorithmic "scientific method" seems able to transcend the basic strategy of trial and error—not if we value new discoveries. The cost of innovation seems to be the risk of failure.

Some critics would have us believe that every little slip made by Darwin—or one of his followers—threatens the whole conceptual edifice he helped build. How impoverished is their understanding of science! Errors are integral to science. But with appropriate critical perspectives, we find them. With appropriate evidence, we remedy them. We can discuss all the errors noted here because we have indeed learned from them. And every lesson has helped ultimately to hone and strengthen the towering theory built on Darwin's sure foundation.

Science would never progress without the courage to fail. Every new idea, even if supported by some evidence, risks being wrong. Darwin was a bold theorizer—and a patient collector of factual details. Taking pride in his achievement means also taking pride in his ability to fail on occasion. It may also remind us of the communal structure of science, whereby errors are noticed and remedied, just as other ideas are cross-examined and confirmed. Darwin's legacy ultimately reflects a monumental collective effort. Accordingly, we may justly commemorate Charles Darwin by celebrating his errors.

PART IV

What Counts as Science

13

Science beyond Scientists

A message of alarm arrives from your cousins: What do you know about the science of "fracking"? Fracking is a way to extract oil and gas. It could potentially generate lots of welcome income in their impoverished rural community—while supplying energy domestically. But possibly dangerous chemicals are injected into the earth and collect in waste ponds. Some residents are worrying about contaminated groundwater.[1] It's potentially quite frightening. But also confusing. Your cousins seek your perspective.

Such a scenario seems to epitomize what "scientific literacy" is all about: being able to interpret scientific claims that inform personal and social decision-making (Figure 13.1). How would a typical citizen or consumer approach this case? Probably search online. Wikipedia. Google. Quick, informative, apparently authoritative answers. Maybe worth investing a half hour of effort, at most.

Delving into the Internet, one can easily find many specialized websites describing how fracking works (energytomorrow.org; fracfocus.org; hydraulic-fracturing.com). They are apparently quite frank about safety issues, which they seem to address fully, including with an impressive quote from a former head of the Environmental Protection Agency. Yet from a more informed perspective, one may find that the genuine facts are also mixed with a lot of questionable claims and spurious "evidence." A lot is left out. The incompleteness betrays bias.

The take-home lesson? What the average citizen or consumer likely interprets as sound science, may not be. Ultimately, good science diverges from *what counts as good science* in the public realm. Here, the challenge is being able to distinguish trustworthy science from junk and industry propaganda. Ironically, knowledge of scientific concepts—the primary stuff one learns in school science classes—is of marginal value. One might thus doubt a pervasive principle (the sacred bovine on this occasion) that in fostering scientific literacy, one should focus primarily on the "raw" science itself, while remaining aloof to the cultural politics of science. Functional scientific literacy includes understanding the media contexts through which science is conveyed—and sometimes misconveyed.

FIGURE 13.1 *Science in the news is often relevant to personal and social decision-making.*

Demarcating Science

The problem of "junk science" has become more acute in recent years—or perhaps it just seems so. Recently neurobiologist Don Agin prominently profiled the problem, with cases ranging from fad diets and longevity schemes to images of "one gene, one behavior."[2] Likewise, physician Ben Goldacre critiqued "bad science" for eight years in a column for London's *Guardian* newspaper. In recent books he takes to task commercial claims about cosmetics; nutrition and vitamins; antioxidants; and "detox" treatments.[3] Physicist Robert Park calls it all "voodoo science," from flawed drug studies to reports that electric power lines can cause cancer.[4] Historians Naomi Oreskes and Eric Conway have exposed the politics behind generating doubt about the dangers of second-hand smoke, acid rain, ozone, global warming, and pesticides. A false image of scientific uncertainty has repeatedly prolonged public debates on those issues and delayed prudent action.[5] Former government epidemiologist David Michaels has noted similar problems in the cases of worker safety regulations on asbestos, dyes, vinyl chloride, benzene, hexavalent chromium, and beryllium.[6] Political reporter Chris Mooney has profiled systematic, selective bias in the use of science even at the level of the Office of the US President.[7] From a legal perspective, Thomas McGarity and Wendy Wagner have analyzed how industry and special interests "bend" health science: by suppressing publication of negative results, harassing researchers, and spinning research findings.[8] The problem extends into the courtroom, too, through biased "experts."[9] Accordingly, the American Association for the Advancement of Science now has a special office to support the education of judges on scientific evidence and testimony.[10] Everywhere we turn, it seems, people with particular ideologies or products "conjure" science as a form of authority for their claims.[11] The impressions of science—what counts as science—can eclipse genuine science.

Of course, none of this is new. Champion skeptic Martin Gardner worked tirelessly to debunk many "fads and fallacies in the name of science": dowsing, ancient astronauts, psychokinesis, orgonomy, anthroposophy, Lawsonomy, and more.[12] Delving deeper into history, one encounters the peddlers of "snake oil" remedies. Even earlier there was mesmerism (so-called animal magnetism). Even in 1610 one can find playwright Ben Jonson satirizing a pair of con men who feign competence as adept alchemists. Society has long been haunted by those ready to capitalize on the credulity of others.

So how does one prepare oneself to interpret such cases, as widespread as they seem to be, from fracking to global warming to miracle cures, where non-scientists purport to have scientific authority? For many, the challenge may seem familiar. One can study the pretenses of pseudoscience, creationism, and the like. That is, one might try to neatly distinguish science from nonscience (or pseudo-, junk, voodoo, bad, bent, or bogus science). In this view, all one needs to do is clearly define what makes science *science*. Sorting should be easy with the right criterion.

Philosophers have certainly tried to characterize the boundary of legitimate science—what they call the *demarcation problem*. But defining the edge of science clearly and definitively proves notoriously frustrating. One wants to exhort consumers to simply be "objective" or "rational," to maintain a skeptical attitude, and to heed the evidence.[13] But this counsel, while easily dispensed, is not so easily articulated in practice. Philosophers have tried to identify signature roles for logic, for verifiability, for falsifiability, for progress, and so on—each abandoned in turn. After their many trials and successive failures, philosophers have largely abandoned this project as unrealizable. There is no simple, single criterion that distinguishes science from nonscience.

The Psychology of Belief

But all is not lost. As the case of fracking indicates, what matters ultimately is a practical understanding of how to distinguish reliable claims from unreliable ones. One can bypass the contentious labeling of "science." A common strategy here is to sharpen critical acumen: learn how to judge claims fully on one's own. For example, the American Dietetic Association presents "Ten Red Flags of Junk Science" for diagnosing diet claims.[14] It advises avoiding recommendations that promise a quick fix. Claims that sound too good to be true are usually just that: untrue. Also: dismiss simplistic studies—those that ignore individual or group differences or that are based on a single investigation. Robert Park provides his own list of "The Seven Warning Signs of Bogus Science." For example, beware anecdotal evidence. Don't trust those working in isolation or claiming that the "establishment" is suppressing their results.[15] One website provides a "Science Toolkit" of six questions for evaluating scientific messages. For example, are the views of the scientific community and its confidence in the ideas accurately portrayed? Is a controversy misrepresented or blown out of proportion?[16]

Of course, one might equally heed the observation that we tend to be beguiled by handy short lists. They certainly help sell magazines. People seem drawn to a small set of enumerated tips, rules, "secrets," or principles. The magic numbers are between six and thirteen.[17] Even such lists, then, may be viewed cautiously.

Indeed, we should give due attention to our inherent cognitive tendencies. Even when good information is available, we do not always recognize it. For example, emotions or first impressions can easily trump due reflection.[18] Prior beliefs can shape what we "hear" or how we interpret it.[19] We can endorse testimony we want to hear (or that just "seems right"). We can discount evidence that doesn't match our previous way of thinking. Even otherwise intelligent people can believe weird things.[20] This is how our minds work. Sometimes they can lead us astray. What counts as science can be victim to how our brains typically function.

That is, we are favorably disposed to some claims, regardless of the evidence, heedless of whether it is accurate or distorted. Science journalist David Freedman describes how we respond to *resonant* advice.[21] For example, we prefer information that is presented as clear-cut and definitive: Why fuss with uncertainties? We prefer a rule that can be applied universally: Why learn more than one? We like things simple: Why bother with time-consuming complexities? We follow others: Why work harder than you need to? These all reflect a pattern of "cognitive economizing." Mental shortcuts are the norm. We also respond more favorably to positive or upbeat pronouncements. We prefer concrete, actionable advice, not merely information or perspective. Drama stirs the emotions. As does novelty. Stories make the facts more vivid. The *appeal* of a claim can be quite strong, apart from the quality of the evidence and mostly apart from conscious deliberation. A good deal of what counts as science reflects the psychology of belief and persuasion more than anything about our understanding of science or evidence.

The profound lesson is that we may not truly engage with evidence, even when it is presented to us. Therefore, we need to see our minds as cognitive machines that are not perfect. Our cognitive dispositions can lead us astray. Understanding the psychology of belief matters. Learning how our minds work—and how they can fail us—is a first step toward securing reliable knowledge.[22] Without proper habits of reflection and self-analysis, scientific evidence will have little import. Ironically, basic lessons in psychology and cognitive science that are essential for interpreting scientific claims are typically not found in the standard K–12 school science curriculum.

Credibility

Another approach to improving the status of what counts publicly as science is to develop the independent ability to reason scientifically. The goal is for citizens or consumers to be able to interpret evidence on their own. So, one may teach skills in the critical analysis of evidence, from recognizing the need for controls or randomized clinical trials to distinguishing cause from correlation or interpreting the degrees of uncertainty conveyed by statistics. Such skills can prove useful, of course (if one is first aware of the common cognitive pitfalls noted above).

Yet the fracking case and others similar to it are again very informative. One may evaluate the evidence if it is available. But a chief problem is that much relevant information is missing. For example, the list of the fracking chemicals injected into the ground is not fully disclosed, under appeals to proprietary information. There are few details about the storage, transport, or treatment of the chemical waste. Also, the geological knowledge required to interpret, say, claims about increased risks of earthquakes is well beyond the average citizen. Even being able to list fully the relevant environmental or health concerns requires sophisticated knowledge of the process. In all these ways and more, interpreting this case requires specialized content expertise, not just generic scientific judgment. Indeed, this seems true for most contemporary socioscientific issues, whether it is assessing the risks in prostate cancer screening or estimating the sustainability of cod fishing off the New England shore.[23] The vision of transforming everyone into an independent scientific agent for all occasions is utopian. In our current culture, we all rely on experts for their specialized knowledge.[24]

Indeed, the impression that one's own judgment can substitute for scientific expertise opens the way to significant mischief. This is the tactic of many anti-climate-change websites (for example, globalwarminghoax.com, globalclimates-cam.com, climatechangedispatch.com, globalwarming.org, and co2science.org). Their effectiveness depends on an individual's sense of autonomy. They encourage independence and the freedom to disagree with the expert scientific consensus. Using fragments of contrary evidence and an intuitive appeal to the concept of falsification, they leverage doubt and disbelief. Of course, their selective use of evidence leads to biased assessments. But without the relevant knowledge, one is unaware that the evidence is incomplete or unbalanced. The uninformed reader is unable to discern which reported evidence is truly reliable. That requires an expert. Pretending otherwise corrupts science.

The approach that encourages individuals to make critical scientific (rather than value) judgments on their own tends to erode good science. In the case of global warming, such advice amounts to implicit dismissal of the professional expertise of the Intergovernmental Panel on Climate Change and of all the scientists who have contributed to its consensus. In today's world of specialized knowledge, a skeptical attitude or disrespect toward legitimate scientific expertise amounts to being anti-science. The challenge, ultimately, is less knowing *what* to trust than knowing *who* to trust. For most socioscientific issues, we need understand not what makes *evidence* credible so much as what makes *testimony* credible. Who are the experts, and why? What is the foundation for expertise? How does one know when someone else can evaluate the evidence effectively? When can one trust their specialized knowledge or judgment? In our world of distributed technical knowledge, understanding expertise and credibility is indispensable to full scientific literacy.[25]

The principles for what counts as science in the public sphere thus differ strikingly from conceptualizing science itself. Understanding how science works internally is not sufficient for interpreting reports of scientific claims in the public media.

One must be familiar with how scientific information flows through the culture and how it is filtered, shaped, and recast as it goes. That is the lesson of imagining a consumer's research on fracking. Of course, even scientists depend on other scientists for their particular expertise. Trust is common. Trust is inevitable. The central challenge, then, is articulating the structure of trust and the tools for assessing any one person's credibility—the focus in the next essay.

14

Skepticism and the Architecture of Trust

Consider the controversy, not long ago, over prostate cancer screening. A presidential task force scaled back recommended testing. But many doctors, citing important cases where screening detected cancer early, disagreed.[1] Whose judgment should we trust?

In New England, fish populations are threatened, according to experts. They suggest discontinuing cod fishing. But the fishermen report no decrease in their catches and defend their livelihood.[2] Whose expertise should prevail: the scientists with their sampling and its inherent uncertainties, or the fishermen with their intimate local knowledge?

There is a lot of alarm about global warming. But maybe it's all "hot air." Many political leaders cite scientific experts who say that the problem is overblown, and just politicized by biased environmental activists.[3] Whose pronouncements should we heed?

As illustrated in these cases, interpreting science in policy and personal decision-making poses important challenges. But being able to gather all the relevant evidence, gauge whether it is complete, and evaluate its quality is well beyond the average consumer of science. Inevitably, we all rely on scientific experts. The primary problem is *not* assessing the evidence, but *knowing who to trust* (essay 13).

In standard lore, science educators are responsible for nurturing a sense of skepticism. We want to empower students to guard themselves against health scams, pseudoscientific nonsense, and unjustified reassurances about environmental or worker safety. But one may want to challenge this sacred bovine. Skepticism tends to erode belief. Blind doubt itself does not yield reliable knowledge. The aim, rather, as exemplified in the cases above, is to know where to place our trust.

The Conundrum of Credibility

The problem of knowing who to trust is not new. In the late 1600s, Robert Boyle reflected on how to structure a scientific community, the emerging Royal Society of England. Investigators would need to share their findings. But reporting added a new layer between observations and knowledge. While ideally everyone might reproduce everyone else's experiments, such redundancy wasted time and resources. Scientific knowledge would grow only if you could trust what others said. But what warranted such trust? Reliable testimony became a new problem.[4]

For Boyle, it was a social problem. You could trust a fellow gentleman, bound to honor and honesty by the social norms of the upper class. By contrast, one could not place as much confidence in a servant or paid assistant, whose private interests might eclipse the pursuit of truth. Accordingly, early Western science became an elite institution, limited to "gentlemen."

The problem in modern science is not so different, although the system has changed. Indeed, as knowledge has become more specialized, the problem has been amplified. We actually know very little on our own. You read a book or newspaper; you watch a TV documentary or webcast; you listen to a friend or a teacher: most knowledge comes from other persons. As noted by philosopher John Hardwig, we are *epistemically dependent* on others.[5] Trust is essential (Figure 14.1).

Indeed, a lack of trust has its costs. According to one sociological analysis, one lab lost the race to discover the structure of thyrotropin-releasing factor (TRF) because of its habit of doubt.[6] Roger Guillemen's group tended to question and redo the experiments performed by the rival lab of Andrew Schally. That cost them extra time. Schally, on the other hand, opted not to second-guess Guillemen's results, but rather to build on them. That allowed him to leapfrog to the conclusion that TRF was not composed exclusively of amino acids. His lab was thus able to identify the other components sooner. It was the first to announce the complete structure of TRF. Trust is integral to scientific progress.

FIGURE 14.1 *Sometimes, trust is essential.*

However, this fact alone does not tell us *how* to exercise trust. Scientific experts, at least, are well positioned to recognize other experts. They can easily use their own knowledge to gauge whether others have the same knowledge.[7] Unfortunately, that's not possible for nonexperts. And therein lies a deep puzzle: how can you identify an expert if you are not an expert yourself?[8]

The problem is illustrated at the popular educational website Understanding Science. In trying to help students untangle media messages, it provides a toolkit for evaluating scientific claims. Its six probes include the following questions:

- Are the views of the scientific community accurately portrayed?
- Is the scientific community's confidence in the ideas accurately portrayed?
- Is a controversy misrepresented or blown out of proportion?

These comparisons can indeed indicate problematic bias. However, it is these very questions that the nonexpert, as an outsider, is unable to answer. Even knowing a bit of the nature of science or how science works cannot help. The consumer of science might therefore seem helpless, susceptible to the whims of whoever *claims* to be an expert.

Of course, we address this same problem in our daily lives. Who is a trustworthy auto mechanic? Who is a qualified doctor or dentist? Yes, even which movie reviewer can you trust to consistently pick your favorites? In these familiar examples, we evaluate evidence, but evidence of a very particular kind. We look for *social data*. Information that reveals someone's performance or abilities. What is his or her experience and demonstrated competence? For a consumer of science, the aim is only slightly different. We do not want the expert's individual "opinion." We seek someone to report and possibly explain the scientific evidence and consensus. Who is a qualified spokesperson for the specific scientific field?

For assessing a building contractor, caterer, or craftsperson, one may seek samples of their work. Online sales also pose parallel problems of trust. Can you have confidence in a seller on eBay whom you have never met? One mechanism for fostering trust is through a system that allows earlier buyers to rate each seller publicly. This generates *a track record* (see essay 9). This concept certainly applies in science. Researchers develop reputations based on their past work. These establish their *credibility* in addressing new cases. Such measures are not absolute guarantees, of course. But they lend confidence. With time, too, we learn the occasional pitfalls: for example, how online sellers can game the system by developing an inflated track record—say, through plagiarism or commissioned reviewers.[9]

Evidence of past performance is not always available, however. So we resort to more-indirect indicators. In our daily lives, if we cannot judge someone's expertise directly, we turn to someone else we already trust—perhaps a partial expert—to provide a "testimonial." That is, we ask for references. Such information is secondary, of course. But it can be valuable, so long as one remains aware of the indirect nature of the evidence and the potential for deception.

Often we rely on venerable institutions to make these assessments of credibility for us. We look for licensed or certified professionals. In science, one looks for appropriate *credentials*—an advanced research degree, publication in rigorous journals, employment at a prestigious institution, service on expert commissions, and so on.

One disregards the need for credentials at one's peril. For example, in 1986 politician Lyndon LaRouche falsely depicted AIDS as easily contagious, transmissable through coughing or sneezing. Despite having no scientific credentials, he was able to persuade over two million voters to endorse mandatory HIV testing and the quarantine of anyone who tested positive.[10] In 1986, unassuming citizen Joe Newman testified before the US Congress about his "energy machine," which he claimed could create more energy than it used. Would that Congress had heeded the federal judge who presided over Newman's earlier patent application. The judge, at least (not a scientist himself), had done his homework. He consulted the National Bureau of Standards, which duly assured him that Newman had not upset the well-established law of the conservation of energy.[11] Credentials, of course, can themselves be bogus. Medical journalist Ben Goldacre takes particular aim at "nutritionists" and other self-appointed health gurus who seem to flaunt all kinds of titles and impressive-sounding references. As a demonstration, Goldacre secured for his dead cat the title of "certified professional member" of the American Association of Nutritional Consultants. Yes, his dead cat. Although it cost him $60. Including the certificate.[12] Credentials are no absolute guarantee. But it is rare that one can vouch reliably for scientific claims *without* such institutionally documented expertise. That can be a first criterion for the nonexpert in ascertaining who to trust in reporting evidence or conclusions.

All these methods are indirect. Their reliability is fragile. So to guard against a single misleading indicator, one may look at multiple indicators simultaneously: do independent assessments concur? In the same way, researchers try to build confidence in understanding cryptic phenomena by using different forms of observation. Agreement among contrasting approaches provides *robustness,* another standard strategy for bolstering confidence in conclusions.

Through such strategies, one may gain confidence that the *source of information* provides reliable, relevant, and complete evidence. Only then might one begin to evaluate the claims themselves.

By Proxy: Credentials versus Experience

A track record, a reputation among professional peers, recommendations by other known experts, and institutional credentials can all be important benchmarks for the nonscientist in assessing someone's credibility on behalf of science. At the same time, these evaluations are, as noted above, indirect. They are *proxies* for gauging the relevant experience (or knowledge, competence, or expertise). Keeping in mind the potential for misalignment between a proxy and the unmediated evidence is important for interpreting exceptional cases.

For example, some scientists present themselves as experts outside their particular fields of expertise. In these instances, they are not really experts at all. A nuclear physicist is no authority on acid rain. Such cases might remind one of celebrities endorsing commercial products unrelated to their actual achievements. We transfer mere impressions from one to the other. It's how our minds tend to work—unless we train them to think more slowly and deeply.[13]

The tactic of using scientists as authorities in illegitimate contexts was adopted by the tobacco industry in denying the adverse effects of smoking on health. One tobacco company enlisted Frederick Seitz to advise its research institute. Seitz had worked on the atomic bomb, advised NATO, and served as president of the National Academy of Sciences and of Rockefeller University. Impressive credentials, indeed. But Seitz was a physicist, an expert on metals and solid state physics. He was not an expert on smoking and health. Politically, though, he harbored some resentments against government interference and saw environmental regulations as trying to thwart democratic freedoms. His "skeptical" attitude and support of "independent" tobacco research were guided by ideology more than by scientific perspectives. Nor was Seitz an expert on several other issues where he flexed his authority: in criticizing the scientific consensus on acid rain, the ozone hole, and global warming. The same story applies to Fred Singer, another noted physicist, who wrote numerous editorials and articles on environmental issues, repeatedly supporting the tobacco and oil industries.[14] The field of expertise matters, not just a generic "scientific" credential.

That was the problem, too, in New Madrid, Missouri, in late 1990. On December 3, the town awaited a strong earthquake. The schools were closed. The city council had stockpiled water. The National Guard had readied an emergency hospital— saying it was just a routine drill. State residents had bought more than $22 million in new earthquake insurance. All because of a single prediction by Iben Browning. But Browning's degree was in zoology, not seismology. Browning claimed to have predicted several earlier large earthquakes—an impressive track record, if true. But few bothered to check whether that credential was genuine. Eventually, the US Geological Survey, with its collective expertise, denounced the prediction and the method used for making it. But a geophysicist at a local university who was the director of its Center for Earthquake Studies endorsed Browning. Few checked his credentials either. Earlier he had relied on a psychic to predict another earthquake, which never happened. As you might have guessed, despite all the pother, no earthquake rattled New Madrid as direly predicted on that occasion.[15] Credentials matter only if they are *relevant*.

What of the critics of global warming? Many of them cite the Leipzig Declaration, a statement signed by 110 people denying a scientific consensus on the issue and asserting that plain satellite observations showed no climate change. That might seem persuasive, if true. However, some journalists duly investigated the credentials of the signatories. Twenty-five were television weathermen: not experts on long-range climate science. Weather is not climate. Other signers included a dentist, a medical laboratory researcher, a civil engineer, a nuclear physicist, an amateur meteorologist, and an entomologist. Of thirty-three European signers, four could not be located. Another twelve denied having signed the document. After those with irrelevant credentials were whittled away, only twenty remained. Many of these were known to be funded by the oil and fuel industry.[16] Not much expertise there, after all. It turns out the declaration was organized by Fred Singer, certainly no expert himself (see above).

In any event, agreement need not be unanimous to regard it as consensus. The Intergovernmental Panel on Climate Change deserves trust. Politicians who continue to dismiss its conclusions are thus not only ill informed about global warming. They are also ill informed about the very nature of scientific expertise—and thereby present questionable credentials themselves as public leaders.

Similarly, one may well question practicing physicians who second-guess large-scale studies based on their personal experience. Most doctors are not medical researchers. While they may be well situated to interpret and explain research findings, they do not necessarily have the appropriate investigatory and statistical backgrounds to evaluate them. Recently, major national expert panels have revised recommendations for mammograms and prostate cancer screening tests. Numerous doctors have expressed dissent, citing individual cases from their own experience. But their anecdotal knowledge is a poor substitute for the systematic studies addressed by the panels. Expert for one task, the doctors are not necessarily expert for another.

Such generalizations about documenting credibility do not preempt the possibility of expertise among those without conventional credentials. For example, in the mid-1980s, AIDS activists became dissatisfied with the drug approval process and medical research. They wanted a voice at the table. Here, they were willing to work for it. They went to conferences and consulted sympathetic researchers. They learned the medical vocabulary and the clinical trial protocols. They studied the virology, immunology, and biostatistics. They thus became fluent in the experts' discourse.[17] In essence, they *became* experts. Robert Gallo, codiscoverer of HIV, was once hostile to them. Later, he admitted, "It's frightening sometimes how much they know." He even described one of the leaders as "one of the most impressive persons I've ever met in my life, bar none, in any field."[18] Experts now acknowledge the activists as members of the community of AIDS experts even though they do not boast the standard credentials. Activists participate as full voting members of the committees at the National Institutes of Health that guide AIDS drug development, and they participate in Food and Drug Administration advisory meetings. Expertise sometimes comes without the customary credentials.

Expertise can also be found among the indigenous peasant farmers of southern Mexico. One might be disinclined to imagine any sophisticated knowledge among those with somewhat animistic conceptions of maize and its "soul." What is one to make of claims that farmland is "hot" at lower elevations, or that the degree of heat can be modulated by the soil's color, consistency, and rockiness, or by shade and wind? Modern fertilizers are considered "hot" too, and the farmers take care not to "burn" the crops. Their crop yields seem modest by comparison with industrialized agriculture. Yet a full analysis reveals that the modest Oaxacan campesinos have a highly developed, ecologically sustainable system. It also addresses the dynamics of replanting, as well as local trade practices. It accommodates the variability of environmental conditions. Scientifically, the system is quite sophisticated.[19] The same complex sophistication is found in the apparently haphazard wanderings of the pastoralists in the Niger River delta.[20] In these cases, the expertise—richly developed local knowledge—is hardly found in formal scientific credentials.

Thus, in the assessment of the fishing stock in North American seas, one should not peremptorily dismiss the knowledge of local fishermen. As it is, fisheries science wallows in uncertainty. At the very least, we need to reconcile the formal science and crude population modeling with the informal but practical expertise of those close to the subject. This is no simple either-or case, where "scientific" experts easily trump the presumably naive nonscientists. Again, experience can, on some occasions, be found without formal credentials. And sometimes one finds, alternatively, credentials without relevant experience. The alignment is not perfect. Assessing who we should trust for scientific knowledge may thus involve some careful discernment.

The Architecture of Trust

Scientific knowledge traces a long path of transformations, from the original set of observations or measurements to the report of a conclusion reaching a scientific consumer.[21] Trust holds the chain together. Even at the outset, investigators learn when to trust their measuring instruments and recording devices. As the data are assembled, members in a lab or research team trust one another. When a paper is submitted for publication, peer reviewers assess the quality of the interpretations, but they also trust that the raw data and images themselves are presented honestly. Fraud does occur on a few occasions. But if coworkers did not detect it, any overt evidence of misconduct is probably already well buried by the time a paper reaches reviewers. On the occasions when a breach of integrity is ultimately found, it is typically the scientific error that is discovered first. Fraud and error follow similar patterns of detection—usually through stymied efforts to build on the original results.

Not all labs publish papers of the same quality. Through experience (and gossip) scientists develop a sense of the credibility of colleagues. This provides a useful (although not infallible) shortcut for assessing the reliability of new results. More-careful assessment of a study's methodology and reasoning may occur—especially if the conclusions conflict sharply with earlier findings or form the basis for future study. But the tedious scrutiny of a paper is generally a backup. With a few well-identified exceptions, trust, again, is the norm. Scientists may well disagree. When they do, one anticipates that further studies will help resolve uncertainties.

Where the conclusions are especially significant, they may get reported in the media or in policy settings. That is where the consumer begins an encounter with science. Not with the unmediated evidence. One may be tempted to regard reporting as just dissemination, a mere transfer of knowledge. But we should not regard it as transparent. It involves editing and framing—and possible deceit. It is another layer of transformation, requiring another layer of trust.

This is where, finally, all the assessment strategies described above matter most. The citizen must assess the evidence—not the scientific evidence, but the social evidence for credibility. First, can one trust the source of information, whether it is a respected newspaper, an advertisement, a website, a talk show host, or a political candidate? If the source is relatively secure, one can then take the next step "back-ward," to assess the credibility of the expert or person making the claims. Known experts and media with confirmed track records are ideal, of course. But fre-quently we must settle for indirect evidence: testimonials (especially from other experts), credentials (or institutional endorsements), the *relevance* of the creden-tials, or other indicators of experience or competence. Consumers interested in reliable knowledge must find the thread that they can trust. Robust agreement, when available, helps.

One can see the whole system at work in an episode from the early 1990s: the prospective link between electromagnetic fields (EMFs) and cancer.[22] The issue became big news when investigative reporter Paul Brodeur published an article in the *New Yorker* magazine in 1989. A pair of studies had detected an association between childhood leukemia and proximity to high-voltage power lines. Should a reader have found cause for alarm? Brodeur had a notable track record. Earlier he had helped publicize the dangers of asbestos and exposed industry efforts to cover up its risks. His credibility seemed sound. The primary researcher was from the University of North Carolina and was largely confirming an earlier, less rigorous study. That checked out, too. So caution about the health effects of EMFs seemed warranted.

But the original two studies were also vague. They were not clinical studies of causation, only epidemiological studies of correlation. Nor was there any physi-ological understanding of how the effect might occur. The overall strength of the EMFs just seemed too low to be biologically significant. In the ensuing media hype, other experts were at hand to note these qualifications. There was no firm consen-sus, mostly because of insufficient evidence. One would have to accept the status of uncertainty and hedge one's actions based on the possible outcomes. But even that required attending carefully to the combined collection of expert opinions. One expert perspective is not always sufficient where consensus does not yet exist.

For the next several years, Brodeur continued to sound the alarm. He published two books and stirred up a great deal of public sentiment. While his *reporting* might have been responsible, he was not a scientist. Yet his own conclusions seemed to receive inordinate weight. Many concerned parents lobbied for local changes and filed lawsuits for damages, as though the science was already well established. The message of uncertainty and the provisional nature of early studies had certainly not been appreciated.

At the same time, medical researchers initiated many further studies (some tak-ing many years) trying to ascertain the nature of EMFs as a possible carcinogen. Finally, in 1996, the National Research Council reviewed over five hundred studies then available. Here was an independent assessment from a panel of the foremost experts in the field. The consensus? For over thirty types of cancer, no evidence indicated harm from EMFs. A key finding articulated a flaw in the original study. The researchers had used distance from power lines as an easily measurable proxy for the degree of EMF exposure. In retrospect, that proved ill founded. Subsequent investigators had been able to enter the homes and measure the EMFs directly. Ultimately, the trust in the original measurement strategy was misplaced. The scientists had to learn that clearly, just as much as the lay public did. Even credible science, alas, is not always free from error (essay 11).

While the scientific debate is largely resolved, lay concerns about EMFs persist. Websites alarm the unwary. They reference some selected research studies and sell books. Yet they do not exhibit the signs of a credible source of scientific consensus. Conspiracy theories find a home. Warnings about cell phones also recur periodically, although their EMFs are even weaker. They seem a danger. But plausibility is not credibility. Good science can often part ways from what counts as good science in the public eye when the lessons here are not heeded.

Learning about who to trust for scientific knowledge (and why) thus constitutes an important challenge. Skepticism, with its exclusively negative orientation, does not solve the problem. The consumer of science needs to be equipped with an understanding of the nature of expertise and the many indirect ways to gauge it—and how those assessments may be limited and when they can fail.

Context also matters. What circumstances motivate and guide the rendering of the scientific information? One additional factor thus looms over all media presentations: the potential of particular interests, notably profit and power, to bias the report. The next essay explores that essential dimension of trust further.

15

Science Con Artists

Deception abounds in nature. Some species are first-rate con artists. Anglerfish with appendages that mimic squiggling morsels to lure unsuspecting prey. Carnivorous pitcher plants that emit the aroma of rotting flesh and attract flies to their doom. Orchids that resemble female wasps, decoys for male wasp pollinators. Cuttlefish whose color and pattern change with the background as a disguise against predators. Deceptive patterns, smells, or sounds in organisms wonderfully reflect the adaptive response to opportunity.

So, too, in human culture? Humans can take advantage of cultural conditions and deceive others to promote their own interests. So if science receives cultural authority, it should surprise no one that those seeking power or profit might try to mimic it. Indeed, the more authority we give to science, the greater the likelihood of science impostors—and the more sophisticated their deceptive tactics. Cultural anthropologist Chris Toumey likens the process to a magician's illusions. Imitators "conjure" science, he says, "from cheap symbols and ersatz images."[1] It is an apt and vivid label. We could just as easily call them science con artists (Figure 15.1). Liars. Cheats. Predictable opportunists who seek our confidence using a semblance of science.

Many proponents of science endeavor to inform citizens and consumers so that they do not succumb to such wiles. The typical posture, too easily adopted I think, is that simple knowledge of the scientific method or of how to evaluate scientific evidence will suffice. By learning about what defines science and what pseudoscience, one supposedly is empowered to debunk the charades. Yet one may well question this sacred bovine.

A sampling of recent historical cases may indicate how prevalent science con artists are in modern society. They are far more significant than commonly assumed. Science and what counts as science publicly can diverge sharply (essay 13). Specific skills may help one sort experts from nonexperts (essay 14). But con artists use various psychological stratagems to gain trust by bypassing questions of expertise. The informed citizen and consumer needs to understand them. Call it an escalation in the evolutionary response of prey to predator.

Consider, for example, the case of German entrepreneur Matthias Rath. Rath's business was selling vitamin pills.[2] In the United Kingdom, he promoted

FIGURE 15.1 *A science con artist?*

them as a cure for cancer, running newspaper ads criticizing chemotherapy and other treatments. Then he went to a welcome environment in South Africa. In 1999 the government there officially denied that the HIV virus caused AIDS and denounced antiretroviral drugs as harmful. Rath ran ads there, too, this time promoting vitamins as a cure for AIDS. The ads described genuine research that showed that vitamins mildly benefitted HIV-positive patients. But then they grossly overstated the conclusions, also claiming that other treatments were ineffective and that vitamins could remedy AIDS outright. Rath also paid for ads in the *New York Times* and the *Herald Tribune*, which he later referred to as favorable news coverage. Credible voices criticized Rath. But his campaign survived them for almost a decade. It is hard to know how many thousands of people died or suffered as a result. We can only hope to learn how he was able to publicly trump scientific consensus, so that we might counteract future such deceptions.

Similar problems plague public understanding on global warming and climate change. In 1989 the George Marshall Institute, a conservative ideology center, began to confuse public perceptions of the emerging consensus of the then-new Intergovernmental Panel on Climate Change (IPCC). The institute's scientists cherry-picked evidence and presented their findings outside the scientific literature, but media and conservative leaders treated their pronouncements as sound science.[3] No one may be surprised, either, that oil giant Exxon Mobil distributed over $8 million in the early 2000s to forty different organizations that challenged global warming.[4]

One begins to see global warming "dissent" as a well-financed advocacy campaign not related at all to science or uncertainty of evidence.

Not everyone who claims to be a scientist, is. Not everyone who presents "scientific" evidence is honest about that evidence. How, then does the consumer-citizen separate the wheat from the chaff? As these two initial cases might indicate, the first challenge is knowing sources' motives.[5] What are their interests in advancing a claim? Will a lie profit them financially? Will it leverage them more political power? That is, consumers need to understand first how science communication may be shaped by persuasive interests.

The concept of social deception is not foreign to most people. We all have fairly good "BS alarms," I think, in familiar social settings. And we know to be on guard *when the speaker is suspect*. Much hinges on that initial judgment of trust.

Of course, science con artists know this. They are thus ready to conceal their interests. They will hide contexts, such as sources of funding or political affiliations, that may threaten their credibility. Indeed, they will aim to actively dampen the sensitivity of our skeptical antennae and our corresponding debunking abilities. But forewarned is forearmed. So one can be prepared. Several common techniques of science con artists, described below, seem especially worth knowing.

Tactic 1: Style

The first aim of confidence artists is, of course, to develop confidence. They do this in part by embodying a confident aura. They look the part. They're smooth talkers. You feel comfortable in their presence. This judgment is made immediately and emotionally, without any conscious intent. Indeed, it requires effort to monitor this first impression and duly check the presenter's credentials. Charisma; smiles; deep, assured voices; colorful prose; snappy sound bites; and "glittering generalities": all may set us at ease and prime us to extend trust. Andrew Wakefield, the doctor whose flawed research ignited fears that vaccinations cause autism, just comes across as such a *nice guy*. Ideally informed consumers acknowledge the invisible power of these psychological pitfalls and remember to methodically "cross-examine" their emotions later.[6]

Style comes in different forms. One subtle feature can be the "professional" quality of publications and media presentations. For example, the enduring battle between biologists and creationists entered a new phase in 2000 with the emergence of the book *Icons of Evolution*. First, the author's education had been funded by the neoconservative Discovery Institute, home to the political campaign known as "Intelligent Design." With a higher degree in biology, he at least presented the superficial semblance of a scientific credential. Far more important, the volume was slick, with excellent production values. It *looked* like a professional academic book. And people do, alas, judge books by their covers. So people could think that the content must therefore be credible. But it was just plain old creationist rhetoric, with all the usual complaints, innuendoes, and omissions. Then came *The Atlas of Creation* in 2006.[7] Filled with gorgeous, large-format glossy photos of fossils.

Printed on high-quality paper to enhance the clarity and vividness of the colors. Designed to impress. And it did. But it was also filled with creationist tripe. Just like the Creation Museum in Petersburg, Kentucky, which strives to *look* like a professional natural history museum. Only with creationist exhibit captions.

These projects presaged the 2009 film *Darwin: The Voyage That Shook the World*. It had all the appearances of a documentary produced by the Public Broadcasting System or the History Channel. Yet the prominent historians who were interviewed were deceived about the film's producers, and their views were not honestly presented in the final edited film segments. Creationists were falsely presented as scientific experts by using the same interviewing style. But who would have the time or resources to check all that? The film looks good, so one assumes that the makers must have been professional in their research, too. They were not. It was a Creation Ministries International scam, borrowing on Darwin's fame to try to erode it. Now there is a new series of videos flogging "Intelligent Design" from Illustra Media: *The Privileged Planet, Metamorphosis*, and *Darwin's Dilemma*. More subterfuge from the well-financed antievolution Discovery Institute.

The same applies now to many websites. For instance, energyanswered.org (now defunct) described itself as "intended to promote fact-based discussion about energy."[8] It looked very professional: well-organized, with clean graphics. And maps. And video clips. No confusing "ads" in the headers or margins. Yet it was funded by the petroleum industry. A bit of careful review could reveal its selective bias. Likewise for co2science.com, a front for industry propaganda on global warming. And cleanairprogress.org, funded by the petroleum and trucking industries, which has closed down since being exposed as a front group.[9] Commercial interests permeate these websites and others. But their persuasive power requires that the visitor not know or suspect this. More well-funded disinformation—here, from antienvironmental science con artists.

Most people are already well aware of the powerful social role of style. They just tend to see it function more through jeans, sunglasses, hairstyles, brand-name fashions, cell phone apps, cars, and so on. We do not apply those perspectives to the context of science communication. Still, everyday experience offers a fruitful platform for analogy. Style encodes persuasive psychological messages, particularly about who is "in" and can be trusted. The savvy consumer is wise to the game.

Tactic 2: Disguise

Because most people understand, at least informally, the dangers of biased messages, science con artists try to hide their interests or associations. Validation in science has typically been marked by publication in a peer-reviewed journal. So that is their aim. But every symbol of science, it seems, can be corrupted. So industries have fashioned ways to publish their views without exposing themselves to the very scrutiny that makes such publication meaningful.

Some industry associations create their own journals. They have impressive names, such as the *Journal of Regulatory Toxicology and Pharmacology, Science*

Fortnightly, Journal of Physicians and Surgeons, Indoor and Built Environment, and *Tobacco and Health*. The list goes on.[10] But these ersatz journals lack rigorous peer review. They provide only the *appearance* of scientific rigor.

At other times, industries seek "credible" publications through credible authors. Having completed a biased study themselves, they enlist—for "due" compensation, of course—a medical researcher or university academic to serve as the author. It's called *ghostwriting*. It's like plagiarism in reverse. And it is an industry unto itself now. You can hire a ghostwriting company to serve your needs.[11] Many journals are responding by requiring authors to disclose conflicts of interest. But authors can lie, and there is little way to enforce honesty. According to a 2003 study, perhaps one in fifteen medical researchers disclosed potential conflicts of interest.[12] So neither publication itself nor the credentials of the lead author, by themselves, can guarantee trustworthy science. Nor are conflicts of interest typically reported in the press.[13] The savvy consumer must always mindfully monitor the potential for conflict of interest.

One of the greatest ironies in recent science con artistry is the emergence of individuals who purport to debunk "junk science" even as they promote commercial and political interests. They pretend to champion good science. Here, their primary goal is typically not to gain scientific status for some ill-founded claim, but to erode confidence in sound science. They challenge findings about harmful chemicals or environmental dangers, all under the rubric of defending scientific rigor. Since no proof is absolute, it's easy to find and exaggerate holes in any study.

Steve Milloy's *Junk Science Judo*, for example, adopts an entertaining and fun posture to debunking science.[14] But its targets are selective, reflecting an antiregulation agenda (embodied by the ultralibertarian Cato Institute, where he works). He defends DDT and junk food in schools and tries to discredit the EPA and climate change research. All in the name of "good" science—but note, not *balanced* or *fully informed* science. It is a fascinating, albeit disturbing masquerade.

The authors of *It Ain't Necessarily So* also pretend to embody a classic skeptical attitude emblematic of science. But they target only studies that support prudent caution on workplace safety or environmental impact. "It's wise to be somewhat skeptical," they say. Ah, but note what follows: "both about fairy tales and about risk narratives." Why just risk narratives? Irony upon ironies, they deny the significance of conflict of interest: "It makes much more sense to look at what the researcher's methodology is, not where the money is coming from. The message, not the messenger, is what demands analysis." It certainly *sounds* simple and appealing. *Ideally*, yes, conflicts of interest and science cons should not exist. Perhaps these authors fear that you will learn their own political affiliations. While they espouse principles of exposing "the failings of journalism, the perversions of policy, and the weaknesses of science,"[15] they do so selectively, and from a consistent pro-corporate stance that seeks to minimize concerns about worker safety and environmental harms.

Knowing who is an expert and who is a bogus "expert" with a conflict of interest matters very much indeed (essay 14). Disguise is a form a lying. Where trustworthy information is important, dishonesty always matters.

Tactic 3: Exploiting Social Emotions

The manipulative techniques of advertising and public relations are well known. Many of us are thus familiar with the role of emotions in persuasion. Products (or ideas) can be associated with pleasant experiences or combined with clandestine sexual imagery. But some of the most dramatic cases of public dismissal of scientific consensus involve another set of emotions, involving sympathy, social cohesion, and loyalty to the "in-group." Here, judgments of trust and credibility are shaped by social relations and emotions tied to a sense of "belonging."

For example, Rath's criticisms of conventional AIDS therapies (see above) were surely more effective in South Africa due to lingering anticolonialist sentiments. Modern anti-HIV drugs seemed to represent the efforts of outsiders to dominate (or destroy). The public health minister, Dr. Manto Tshabalala-Msimang, could easily convince others that local customs in nutrition, embedded in African culture and history, could effectively combat an "alien" disease.[16] Somehow, the science was upstaged by social identity and cohesion.

Similarly, opposition to the fluoridation of public water in the 1950s and '60s was shaped less by the scientific evidence than by fears of government intrusion. Personal autonomy (or, in the midst of the Cold War, fear of totalitarianism) seemed prior to addressing any health effects from the fluoride. People rallied together to defend themselves. Identity with fellow antifluoridationists often seemed to dictate how individuals would subsequently choose *which* scientific evidence they would trust.[17] Social alliance was the basis for judging scientific reports.

In 1986 charismatic leader Lyndon LaRouche managed to tap into fears of AIDS to persuade many people that it was contagious through proximity and casual contact. Uniting people xenophobically under a perceived shared threat (disease and homosexuality), he persuaded over two million Californians to vote for mandatory HIV testing and quarantine. Again, the social dimension trumped the trustworthy science.[18]

More recently, we have seen concern about autism misdirected at vaccines as a cause. There was only ever one, quite insubstantial scientific study that supported this causal claim, and it has since been retracted. Yet the anxiety in a consolidated core group of parents triggered widespread cultural distrust of the measles vaccine (and others). Vaccination rates in Britain dropped so low that public health officials worried about a significant measles epidemic. One thirteen-year-old boy died—the first death there from measles in fourteen years. Even now, shared anti-vaccine sentiment unites some social networks so strongly that anyone who presents contrary scientific information is prejudged as a likely apologist for the pharmaceutical industry. Again, social connections have proven a basis for judgments about trusting scientific claims.

There seems to be a neurohormonal dimension to this behavior. Recent research has shown that oxytocin released from the hypothalamus promotes empathy, trust, and generosity—at least within a group.[19] At the very same time, it also promotes out-group antagonism and distrust.[20] Trust and in-group sociality seem to be closely related neurophysiologically.

Science con artists can exploit and benefit from these sociopsychological tendencies. First, they often present themselves as "just plain folk"—a strategy to become accepted *socially* as one of the group. The nature of an implied group can be further generated or amplified through eliciting fears of external threats, name-calling, or rhetorical venom.[21] Alleged conspiracies tend to evoke social consolidation, and with it, trust. A cautious, deliberative response by the consumer-citizen will be met with the con artist's statements of the urgency of the situation. There will be discussion of government cover-ups and efforts to suppress the "real" scientific evidence. The allegations effectively divert attention from the scientific literature or discourse of experts.

The social dimension of trust, then, proves relevant in the cultural transfer of scientific information. It helps remind us, perhaps, that chains of trust are just as important as skepticism. One needs to know not how to doubt but precisely where to place trust—or with whom.

Tactic 4: Conjuring Doubt

One especially notable con artist strategy has been adopted with increasing frequency in the past several decades: conjuring public doubt amid scientific consensus.[22] This tactic has proven quite effective for diminishing or delaying policy or regulatory action where scientists have documented potentially harmful practices. The con exploits the popular notion that science is, or should be, certain. Thus if there is any notable dissent whatsoever, the science will not seem "truly scientific." Until the science is certain, so the rhetoric goes, we should not rush to judgment. A cautious person waits "prudently" for all the evidence, "just in case." It sounds so reasonable. Unless one knows that the science is definitive, and someone is profiting while knowingly harming others.

The con tactic, then, is to foster a public image of uncertainty, even where most experts agree or the preponderance of evidence is clear. Psychologically, it seems, an ounce of uncertainty is worth a pound of doubt. Doubt, in turn, can further be reframed rhetorically as "probably wrong." This tactic was pioneered in the late 1960s as the tobacco industry fostered doubt about research on the adverse health effects of smoking. One 1969 industry document referred explicitly to the campaign's intent: "Doubt is our product since it is the best means of competing with the 'body of fact' that exists in the minds of the general public. It is also the means of establishing a controversy."[23] Since that time, the strategy has been deployed repeatedly. In many other cases where researchers had determined a measurable harm, industries tried to persuade the public that the evidence was not definitive (Figure 15.2). Sometimes they used public relations firms to help shape what counted as science in the public realm—for example, Hill & Knowlton; Exponent, Inc.; the Weinberg Group; or ChemRisk. Ironically, those folks do know a lot about science—the science of persuasion.

Conveying the expert consensus on human-caused global warming and climate change to the general public is, of course, another major case, which still seems to

- second-hand smoke

- acid rain

- chlorinated fluorocarbons (CFCs) and the ozone layer

- DDT use outside the U.S.

- formaldehyde

- hexavalent chromium

- vinyl chloride

- lead

- ephedra

- global warming and climate change

FIGURE 15.2 *Some cases of manufactured doubt in science*

haunt us, at least to judge from official statements by some major political figures. This is a prime example of the ongoing significance of conjuring doubt.

How do the con artists conjure doubt? They may question whether animal models are representative of human physiology. One can always *question* this, whether the ultimate criticism is *justified or not*. They may challenge the representativeness of human samples too. They may flat out challenge data as unreliable. They may enlist statisticians to reanalyze published data using modified parameters, in order to reduce the statistical significance. They often find and highlight single exceptions, discounting the overall balance of evidence. They emphasize extraneous causes and possible confounders—vague sources of error that can be *imagined as possible* without having to be documented as actually relevant. If the aim is an *image* of uncertainty, one does not need to win the debate, or even justify one's argument fully. One just needs to provoke sufficient skepticism.

In general, the conjurers of doubt try to prompt others to second-guess the experts. They portray flaws in the consensus as so simple and obvious that even unschooled nonscientists could easily detect them. Con artists capitalize on an individual's sense of autonomy, the belief that one can evaluate all the evidence on one's own. If an ordinary person can understand a shard of counterevidence, and the experts have not heeded it, then the experts must apparently not be so expert. As David Michaels (former Assistant Secretary for Environment, Safety and Health in the Department of Energy) once noted, this is how one begins to replace sound science with what merely "sounds like science."[24]

Tactic 5: Flooding the Media

When all else fails, one can merely generate a public impression of science through "advertising."[25] The information need not be complete. Or responsible. For example, prominent global warming critic Fred Singer arranged to co-author a paper with a climate change expert, (a rather reluctant) Roger Revelle. Although including some key phrases that misrepresented Revelle's views, the paper was published after he died. As one measure of Singer's intent, the paper appeared in a "showcase" journal of an elite social club in Washington, DC. The misleading claims then reappeared in numerous conservative editorials and public remarks (and even in a nationally televised vice presidential debate), all positioned to diminish the impact of Revelle's cautionary claims elsewhere.[26] Science does not have its own centralized institutional voice. One thus needs to be concerned about who presents the scientific "message," where, and how.

Psychologist Daniel Kahneman notes that our minds have certain blind spots. One is that we typically base our judgments on what we have heard and seen, without a care about possibly relevant information we have not yet encountered. "What you see is all there is," he says.[27] Accordingly, we tend to endorse whatever is familiar, whether we are fully informed or not. Without active reflective analysis, we may easily believe the preponderance of public claims.

Those who can underwrite "public awareness" can thereby influence what counts as science in the public realm. They may do so through mass mailing. Or booths at state fairs. Or radio talk shows. Or websites, blogs, and tweets. Or comments by political candidates. *World Climate Review*, a contrarian newsletter funded by the fossil fuel industry, is distributed free to members of the Society of Environmental Journalism.[28] The Heartland Institute, a well-financed advocacy center, sponsors a network of individuals to write letters to the editors of major US newspapers challenging the science that does not match its conservative agenda. More ghostwriting.[29] There is no formal system of accountability or checks and balances in public science communication, as there is in science. So Fred Singer could make further misrepresentations about the 1995 IPCC report in an opinion piece in the *Wall Street Journal*. The editors were entitled to disregard most of the letters from outraged scientists. And others were free to (and did) cite Singer's essay.[30] For many citizens, what you read is all there is. Flooding the media is just another science con tactic.

Counteradaptation

These five tactics for advocacy and fostering trust (and one might surely enumerate more) are hardly limited to science communication. But the context of science is special. By eclipsing the relevance of credible evidence, these various forms of persuasion can misinform public discourse and decision-making.

Dealing with the tactics falls outside the standard textbook view of science as defined by the "scientific method." They thus may seem irrelevant to science proper. But knowing about them seems essential for anyone trying to assess scientific claims *in a practical, cultural setting*. The average person cannot find and judge every bit of evidence on her or his own. One depends on others with expertise. But that leaves one vulnerable to deception. The best protection is to know how to diagnose, detect, and thus also deflect the deception. And then find some real experts (essay 14).

Some people worry about fraud or dishonesty in science. I worry instead about fraud, deception, and misplaced trust in science communication *beyond* the scientific community. As documented above, efforts by nonscientists to mislead others about scientific consensus are widespread and have concrete consequences for the environment and human health.

Some people worry about a conflict between science and religion. I worry instead about the conflicts between science and power and between science and the blind drive for profit. These are the forces in modern society most likely to corrupt good science and science communication.

Some people worry about pseudoscience and ill-informed views about the nature of scientific evidence. I worry instead about how *what counts as science* in our culture diverges from the actual science. I worry, too, about the wide range of tactics for shaping science communication and fostering trust in hollow "scientific" claims.

In short, it seems, our modern cultural condition warrants substantially more attention to science con artists. Recall again the smokescreen claim by a group of conservative science "critics" (noted earlier): "It makes much more sense to look at what the researcher's methodology is, not where the money is coming from. The message, not the messenger, is what demands analysis." That is precisely what the science con artists hope everyone will believe. Messengers can deceive. They can be confidence artists. One needs to assess the messenger before heeding any message. But will we thwart their deceptions by developing a culture with the appropriate countertactical skills?

PART V

Naturalizing Cultural Values

16

Male, Female, and/or —?

Intersex individuals are coming out of the closet. Witness, for example, the 2003 Pulitzer Prize in Fiction for Jeffrey Eugenides's *Middlesex*.[1] The story follows someone with 5-alpha-reductase deficiency, or late-onset virilization.[2] Imagine yourself raised as a girl, discovering at puberty (through cryptic, piecemeal clues) that you are male instead. Or male *also*? Or male *only now*? Or "just" newly virile? The condition confounds the conventionally strict dichotomy between male and female, masculine and feminine. It teases a culture preoccupied with gender.

What are male and female, biologically? How does nature define the sexes, and sex itself? The questions seem simple enough. Seeking answers, however, may yield unexpected lessons—about the role of biological definitions; about assumptions concerning universals, rarities, and "normality"; and about the power of mistaken conceptions of nature to shape culture.

Sorting the Sexes

Conceptualizing sex as male and female seems straightforward. In the standard version (familiar even to those unschooled in biology), females have two X chromosomes, while males have an X and a Y. They have different gametes: one moves, one stays stationary. These differences seem foundational. They seem to explain why male and female organisms have contrasting gonads, contrasting hormone-mediated physiologies, and contrasting secondary sex characteristics. Once-homologous organs follow divergent developmental trajectories. Perhaps even contrasting behaviors express the purported evolutionary imperative of each gamete: the "promiscuous," uncaring male of cheap sperm, and the cunning, protective female of big-investment eggs. The apparent alignment of the two sexes through all levels of biological organization seems to validate this categorization as scientifically sound.

Good biologists know better. First, sex may be determined in many ways. Birds use a "reversed," WZ system, where females have the distinctive chromosome. Many insects have a haplodiploid system, where sex is determined by having a single or double set of all the chromosomes. Crocodiles and turtles develop their sex in response to temperature cues, not genes alone. The spoonworm *Bonellia* responds instead to whether females are absent or already present in the area.[3] In humans, widely documented chromosomal rearrangements (XO in Turner's syndrome and XXY in Klinefelter's syndrome) are further exceptions to the conventional formula and yield individuals with a distinctive mixture of traits not conventionally associated with either male or female.

More challenging puzzles arise with XY *females* and XX *males*. In these rare human cases (which occur about once in twenty thousand births), segments of the X and Y chromosomes are exchanged when the sex cells divide. The Y chromosome may lose or the X chromosome gain the critical *Sry* gene, which regulates many other "male"-related genes.[4] Of course, such exceptions only seem to affirm another presumed, more fundamental reality: identity as fixed by genes (essay 19).

The fascinating cases continue, challenging our unquestioned commonplaces—our sacred bovines. The world is so incredible. Why stuff it into pigeonholes? Indeed, we must guard against the label "scientifically known" ironically discouraging any further scientific inquiry. In intersex humans (as portrayed in *Middlesex*), hormonal levels, ineffective hormone receptors, or developmental branches lead to mosaics of sexual characters. For example, external genitalia may contrast with internal gonads. Bodies may exhibit almost any combination of "male" and "female" parts, mixing different elements of the sexual anatomy, including position of the gonads, urinary plumbing, size of breasts, facial hair, hair loss, invaginations and protuberances, ejaculates and menses, and vocal timbre.[5] All these traits do not sort neatly according to chromosomes *or* gonad type. Some "mixed" patterns are actually typical within other mammal species, from spotted hyenas and kangaroo rats to bushbabies and Old World moles.[6] Male and female, one might learn, are suites of characters, not essential, comprehensive dichotomies. Imagine the problem for the intersex individual: Which box do you check on a passport application?

Many fish, including wrasses, parrotfish, and groupers (and others found in tropical aquaria), may *change* sex during their lifetimes.[7] In the cleaner wrasse, for example, a community typically has many females and one male which releases a pheromone that inhibits others from developing as males. When the male dies, the largest female begins changing sex in a matter of hours to become the new male. In other species, such as the clownfish, males later become female (a detail not included in the popular film, *Finding Nemo*) (Figure 16.1). Some gobies go even further, changing sex more than once. Organisms develop. Sexed anatomies and physiologies may change. Sex, it seems, need not be viewed as a predetermined or fixed identity. How informative might that knowledge be to an adolescent experiencing an emerging sexual body?

Of course, organisms need not be just male *or* female. Some are male *and* female simultaneously. Hermaphroditism is found in snails (Figure 16.2), earthworms, barnacles, and many deep-sea fishes.[8] Most plants, too, have both male and female organs. Some mammals develop *ovotesticular* tissue or, rarely, both an ovary and a testis simultaneously (although none seem to reproduce with both gamete types).[9] Similarly, the steroid hormones that contribute to various male- and female-related traits, such as facial hair and breasts—once popularly envisioned as mutually

FIGURE 16.1 *Male clownfish later become female—a fact not divulged in the popular animated film* Finding Nemo.

exclusive "sex" hormones—are produced by male and female alike.[10] Are we too accustomed to binary, either-or categories?

Finally, one may encounter problems sorting reproductive behaviors. The whip-tail lizard is an often-cited example: females engage in pseudocopulation, critically stimulating development without a fertilized egg.[11] Another case involves the *fruit-less* gene in fruit flies. Different mRNA transcripts of this one gene are spliced to produce three proteins in males but not in females. Those proteins contribute to

FIGURE 16.2 *Snails are hermaphrodites: both male* and *female.*

developing a circuit of sensory cells that detect odor signals and so control mating behavior. Recently, in a dramatic experiment, females were genetically manipulated to splice the mRNAs. They exhibited mating behavior toward other females. By contrast, males whose splicing was inhibited barely courted.[12] Behaviors, too, are hard to characterize as necessarily exclusively male or exclusively female.

Not long ago, Harvard political philosopher Harvey Mansfield characterized what he called "manly men."[13] Ironically, his phrase is far from redundant. Indeed, it underscores how behavior and sexed anatomies need *not* align according to any pattern. Remarkably, he nonetheless appealed to science to label certain behaviors as "manly," or associated with only one sexed body: behaviors he presumably values for himself. In many species—for example, seahorses, some pipefish, and aquatic birds such as the wattled jacana, Wilson's phalarope, and spotted sandpiper—males are the primary protectors and nurturers of offspring.[14] Does one thus call these males maternal? Or paternal? Or just parental? Mansfield's mere use of the term "manliness" bristles with biological contradictions.[15]

All these cases exemplify one virtue of comparative biology. They help us perceive who we are by showing us how organisms can be organized otherwise. Sex is not a permanent identity. Sex is not solely genetic or inherited. Sex is not either-or. The characterization of male versus female is ultimately not so clear-cut as the simple labels might indicate.

In biology, exceptions abound. Absolutes are few. Here, no "one body, one sex" rule holds universally. No single-trait definition applies to all other traits uniformly. An excursion into sorting male and female, then, more than highlighting mere curiosities, can show how some of our most basic assumptions about nature are mistaken. We may also begin learning to check the all-too-easy tendency to oversimplify nature.

Domesticating Nature

Aristotle surely expressed an ideal when he advocated adopting categories that would "cut nature at her joints." That goal assumes, of course, that nature has clearly defined joints. Yet in nature, sexed traits do not sort uniformly. Intersex mosaics may combine traits that we conventionally designate as *either* male *or* female. Traits at different levels of organization do not always align according to the popularly imagined dichotomy. Why, then, did the sharp dichotomy emerge, and why does it persist?

Consider two models for human sexual development.[16] In one, the body begins as indeterminate—unsexed, perhaps. Each embryo is then assumed to respond to hormonal (or genetic) "switches" that trigger only one of two diverging trajectories, male *or* female. In this model, intersex individuals are strange hybrids or deviants. They may even seem "unnatural" monsters (essay 17). Note that the possibilities this model portrays as scientifically "normal" echo familiar cultural categories.

In the other model, the original mass of cells is multipotentialed. In a sense, the embryo is *both* male *and* female, because no developmental opportunities have yet been closed off. As the organism develops—both before and after birth—in response to local and temporal cues, some potentials are followed, while others are not (sometimes they are even lost). A constellation of "male" and "female" traits emerges. In some cases, certain unexpressed potentials might later be pursued, as in late-onset virilization, or perhaps even restored, as in intentional surgical or hormonal intervention. In this second model, indigo hamlets a type of reef fish that within just a few hours may switch several times from delivering sperm to having their own eggs fertilized, do not change bodily identity at all. Rather, they merely exhibit physiological options—here, related to alternative reproductive functions.

Thus, at least two models of sex are possible:

(1) male *or* female, or
(2) male *and* female, each in differing degrees.

Why then, given hermaphrodites and intersex conditions, do we typically interpret the duality as "or" rather than "and"? Are we subtly influenced by cultural frameworks—for example, based on the either-or competitive rhetoric that pervades athletics, economics, and academic assessments (essay 8)?

Whether one envisions sex as male *or* female or as male *and* female, at least the options seem reduced to two. Biologically, that seems to reflect the very nature of sexual reproduction. Males mate with females! What would seem more "natural" as a benchmark? Sex seems plainly *binary*.

Comparative biology, once again, may upset the conventional order. Looking beyond humans, one can find sexual reproduction without dimorphic sexes. Moreover, some species may have more than two sexes or mating types. All these systems nevertheless express the fundamental feature of sex: *genetic recombination*.

For example, gametes need not be egg or sperm. In the algae *Chlamydymonas* and *Ulva*, the gametes are undifferentiated. Reproduction is nonetheless sexual. The special cell division to produce sex cells does occur. Gametes do fuse. But *both* gametes have flagella and move to find each other. There are no small gametes and big gametes, hence no identifiable male or female. Other algae, fungi, and protozoa exhibit the same pattern.[17] Sexual reproduction occurs widely *without sexes.*

Further, sexual reproduction recombines genetic material only when different types meet: sex alone is no guarantee. Accordingly, further mechanisms may help foster outcrossing or ensure the critical shuffling of genes. For example, in many plants, the pollen (male) does not germinate unless the host stigma (female) is of a different genetic type, determined by specific sterility alleles. Mating strains may regulate inbreeding and promote outcrossing. *Tetrahymena thermophila*, a protozoan, has seven mating strains. *Schizophyllum commune*, a mushroom, has over twenty-eight thousand.[18] If one interprets sex as who can mate with whom for the purpose of recombining genes, then there can be a large number of sexes indeed!

Mating strains typically differ chemically. However, anatomical differences of mating types were also noted by none other than Charles Darwin. Darwin discussed *heterostyly*, where the lengths of the female styles (even in some common garden flowers) vary in discrete forms, long and short.[19] Male stamens, too, vary in length. The relative positions of the sexual structures dictate how pollinators transfer pollen most effectively between different flowers. Hybrids are more successful reproducing, Darwin found. In some cases Darwin noted *three* discrete lengths (Figure 16.3), leading to what he profiled as three pairs of males and females, each akin to reproductively separate species. "No little discovery of mine ever gave me so much pleasure as the making out the meaning of heterostyled flowers," Darwin

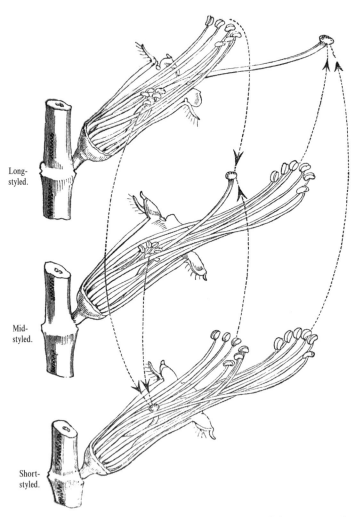

Long-styled.

Mid-styled.

Short-styled.

FIGURE 16.3 *Heterostyly in common flowers: different positions of the stigmas and anthers ensure cross-fertilization between types.*

noted proudly in his autobiography.[20] Darwin, of course, well appreciated the evolutionary significance: "We may feel sure that plants have been rendered heterostyled to ensure cross-fertilisation."[21] Functionally, style type seems as important to sexual reproduction as sex itself.

Other biological systems, like heterostyly, also involve discrete reproductive forms. Even gametes may exhibit more than two distinct types.[22] Nor is sexual dimorphism (differences in male and female anatomies) a universal rule. One especially striking system was recently discovered among two ant species of the genus *Pogonomyrmex*. Independently of male and female, the ants have two distinct mating types. A queen that mates with a male of her own type produces more queens, essential to the continuity of the colony. However, she must also mate with a male of the alternative type to produce workers, who keep the colony functioning. *Both* matings are necessary for the *colony* to survive, although each *individual* ant has only two parents. Joel D. Parker argues that these ants thus have *four* sexes.[23] For effective reproduction at the level of the colony, *three* sexes—one female and *two* male—seem essential. These ants certainly illustrate the complexity of reproductive systems and mating forms. Sex does not reduce to "simple" mating between dimorphic males and females.

Intersexes, null sexes, mating types, multiply morphed sexes: one may be tempted to dismiss them all as merely exceptions. Or as insignificant because they are rare. Here, one may easily confuse "normality" as numerical frequency with "normality" as a presumed *value*. Indeed, the "exceptions" are not so rare as that label might indicate. Intersex humans occur probably about one to three times in every two thousand births (more, if one is liberal in one's interpretation).[24] (For comparison, the incidence of cystic fibrosis is about one in two thousand, and that of Down's syndrome about one in eight hundred.) As noted, among plants hermaphroditism is more common than not. Of course, prior conceptions define what is an "exception." "Normal" is a cultural judgment (also see essay 17). Ultimately, the relative frequencies are unimportant. The diversity of examples is significant because it challenges the very concept of sex. Viewing sex as dichotomous or as uniformly male-or-female, one cannot fully characterize—or fully appreciate—nature.

Reconceptualizing sex is profoundly challenging, because the male/female dichotomy seems such a plain biological fact. Far more deeply, however, it permeates our cultural organization, from names and dress to military conscription, career and athletic opportunities, and various other enfranchisements—even to toys and games. The temptation among many may surely be to cast intersex organisms and other complex cases as *unnatural*—that is, as violating some "scientific" notion of male and female. The inescapable irony, of course, is that all the cases discussed above are products of nature. The "unnatural" epithet uses a culturally laden concept of nature to incorrectly interpret nature itself. Namely, we tend to *naturalize* the cultural notions of male and female—and then validate them as "objective" in the name of science.[25] Exploring biological notions of male and female may thus yield insights about the nature of science. The challenge is to interpret how observations and cultural perspectives interact in making and using such scientific concepts.

Delving into the biological spectrum of sexed bodies, behaviors, and systems of sexual reproduction is unabashedly sensationalistic. But it can also lead to genuine and profound lessons. The significance of that understanding may perhaps best be measured by reflecting on why we pose, and how we choose to answer, the ubiquitous—and presumably unambiguous—first question about newborns: "Is it a boy or a girl?"

17

Monsters and the Tyranny of Normality

In the sixteenth and seventeenth centuries, monsters were wonders (essay 1). Anomalous forms—like conjoined twins, hermaphrodites (essay 16), hydrocephalic babies, or the extraordinarily hairy Petrus Gonsalus and his equally hairy children—amazed people. They evoked a spirit of inquiry that helped fuel the emergence of modern science. Today, however, such bodies tend to strike us as freakish or grotesque—possibly even "against nature." How did our cultural perspective, and with it, our values and emotional responses, change so radically?

The shift in cultural views, ironically, paralleled deepening scientific understanding. Exceptions and anomalies can be powerful investigative tools. In this case, human monsters eventually prompted a new science, *teratology*, which compared normal and abnormal development. The scientific explanations and categories seemed to support value judgments. The history of monsters helps reveal the roots of a common belief (another sacred bovine): that the "normal" course of events reflects nature's fundamental order. Well construed, monsters can help us rethink the meanings of normality and of the concept of laws of nature.

Leveraging Exceptions

Monsters are fascinating, of course, because they do not fit customary expectations. Such exceptions can be valuable opportunities for interpreting the *un*exceptional.[1] One can begin to look for the relevant differences that reflect the underlying cause in both cases. It is a classic research strategy, especially in biology. Loss or modification of a structure can highlight its function.

So, for example, vitamins were discovered through vitamin deficiency diseases, such as scurvy and beriberi. Likewise, the role of proteins in gene expression emerged from studying heritable enzyme deficiencies, such as alkaptonuria and phenylketonuria. Sickle cell anemia has become a classic example for learning in part because it was important historically in understanding hemoglobin and protein structure as well as the evolutionary consequences of the multiple effects of a single gene.[2]

Similarly, diabetes provides insight into the physiology of regulating blood glucose and the hormone insulin. Slips of the tongue are clues to how the brain processes language (missed notes in playing piano, too!).[3] Autism is opening understanding of the physical architecture of nerve cell signaling.[4] The dramatic influence of nonnative species, such as zebra mussels or gypsy moths, reveals how coevolution stabilizes ecological relationships. Biological functions at all levels can be deciphered when customary patterns are disrupted.

Puzzling exceptions, or anomalies, were an obsession in the early 1600s for Federico Cesi, founder of perhaps the earliest scientific institution, the Accademia dei Lincei.[5] Cesi's foremost goal was to document and classify every organism on earth. He was thus intrigued by specimens that fit two categories at once. A bat seemed like both rodent and bird. But it could not be both. It posed a puzzle for how to adjust the existing categories. What was the essential nature of a goose barnacle? Was it a fungus that generated a shell? A mollusk that produced feathers? Embryonic birds? Something else? In the same way, Cesi approached double fruits, lemons with various excrescences, and other "monstrous" forms as cryptic clues to nature's order.

Physician and anatomist William Hunter would echo such sentiments nearly two centuries later: "Even monsters, and all uncommon, and all diseased animal productions, are useful in anatomical enquiries; as the mechanism, or texture, which is concealed in the ordinary fashion of parts, may be obvious in a preternatural composition." In such examples, he rhapsodized, nature "has hung out a train of lights that guide us through her labyrinth."[6] Monsters offered deeper insight into ordinary nature.

From Anomalous to Pathological

The growing desire in the 1600s to understand monsters, along with other wonders and preternatural phenomena, helped motivate the emergence of modern scientific investigation. As the study of nature expanded, belief in nature's regularities became more firmly established. Over the century, the concept of "laws" of nature developed—with ultimately profound consequences.

Initially, such prospective laws included how monsters formed. In 1703 Bernard de Fontenelle, secretary of the Paris Royal Academy of Sciences, expressed such faith: "One commonly regards monsters as jests of nature, but philosophers are quite persuaded that nature does not play, that she always inviolably follows the same rules, and that all her works are, so to speak, equally serious. There may be extraordinary ones among them, but not irregular ones; and it is even often the most extraordinary, which give the most opening to discover the general rules which comprehend all of them." In monsters, he professed later, nature "cannot avoid sometimes betraying its secret."[7]

Monsters thus became integral to early eighteenth-century debates about the "laws," or principles, of development. For example, did the mother's imagination contribute to the form of the fetus? Exotic monsters seemed the result of dreams or bizarre experience, not ordinary nature. Such a view was widespread until challenged in 1727 by James Blondel, who helped shift views toward more purely material processes. Also, were monsters foreordained or the result of "accidental" factors during fetal growth? Nicholas Lemery advocated the latter, suggesting that conjoined twins resulted when by chance two eggs merged. His arguments were not uniformly accepted, however. Another debate concerned whether embryos exhibited essential structure from the outset (preformation) or whether their structure developed only gradually, varying with context (epigenesis). For many decades monsters were important test cases for any theory of development.[8]

In the early 1800s, the belief in nature's regularities and patterns in organisms was expressed through "philosophical anatomy." Different organisms, both within and across species, could be related through idealized forms. German Romantic poet and naturalist Johann Wolfgang von Goethe saw unity of structure even within organisms. All plant organs (flowers, bracts, tubers, tendrils, and so on) were, he claimed, fundamentally leaves. In France, Étienne Geoffroy Saint-Hilaire advocated a law of "unity of composition." Geoffroy extended the unity to monsters. "There is monstrosity," he noted, "but not, by virtue of that, suspension of ordinary laws." He called it a "manifest contradiction" to imagine two separate sets of laws governing development. All variant organisms reflected an underlying model pathway. "The normal state of humans may be considered like the abstract being, or generic being, and their different pathological deviations, like the species of this ideal type."[9]

Geoffroy's follower Étienne Serres studied monsters—notably, a famous pair of conjoined twins, Ritta and Christina (Figure 17.1)—and viewed them as resulting from arrested development. Their forms, in Serres's perspective, were incomplete. Geoffroy tried to demonstrate such effects experimentally. Knowing that an egg loses weight as the chick develops, he focused first on the presumed exchange of fluids across the shell. He abraded eggs, pricked them, or covered them with wax or varnish. Later, he considered mechanical disruption. If adherence of embryonic surfaces was altered, he reasoned, changes in blood flow might arrest development in particular locations. Geoffroy varied the position of the eggs, standing some on one end, some on the other. He soaked others in water. Yet others he exposed to periods of lower temperature. That is, he endeavored to mimic and thereby demonstrate the causes of monstrosities. The results were inconclusive. Organisms tended to die rather than develop differently. Geoffroy nonetheless continued to propound his theories. Still, he had opened an important new area to investigation. Monsters were now subject to experimental science.[10]

Geoffroy's son, Isidore Geoffroy Saint-Hilaire, continued his father's work and helped it achieve greater respectability. He consolidated and extended his father's system of classifying human monsters. He also applied the concept that developmental rates of particular body parts might vary (even if not fully "arrested"). While studying primates, Isidore noted that the skulls of the young were relatively larger than the adults'. He speculated (before Darwin had published on evolution) that different developmental rates might explain human brain size. Here, he echoed some of his father's controversial views that monsters might provide clues about

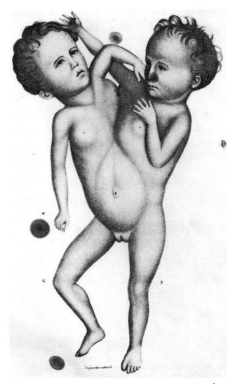

FIGURE 17.1 *The conjoined twins, Ritta and Christina, studied by Étienne Serres.*

how one species transformed into another, as theorized by Jean-Baptiste Lamarck. Most important, perhaps, Isidore gave the new science of monsters a firmer existence by naming it. It is still known as *teratology*.

Isidore stressed that despite their unusual appearance, monsters should not be regarded as "failed" normal beings. But philosophical anatomy and its law-like ideals dominated thought. Teratology became a science of *pathology*. Explaining the normal processes behind the anomalous forms ironically contributed to monsters being cast as "abnormal."

The development of statistics during the 1800s contributed further to viewing monsters negatively. Astronomers and geographers realized that their multiple measurements of the same stars or landmarks varied. The variation exhibited what we now commonly recognize as a bell-shaped statistical distribution. But the stars and land were obviously not moving! So some measurements must be "wrong." The desired figure, or ideal, was surely the average of all of them. Statisticians thus labeled the variation—today's bell curve—as the "Law of Error." Meanwhile, numbers about all kinds of social phenomena were exhibiting the same Law of Error. The statistical regularities were construed as social *laws*. In the 1830s mathematician Adolphe Quetelet suggested that rather than discuss groups and their variability, one could refer instead to the "average man" (*l'homme moyenne*) as representative of the whole. That concept further endowed the average, or common, with value, while peripheralizing others. Statistics thus helped amplify, and apparently justify, value distinctions between the "normal" and deviations from it.

With faith in law-like regularities, philosophical anatomy, teratology, and statistics, monsters changed in the 1800s from anomalous wonders to pathological errors. Consider, for example, the case of Joseph Merrick, also known as the "Elephant Man," in midcentury (Figure 17.2). Merrick exhibited Proteus syndrome (genetically based excessive bone growth). His head was enormous and bulbous; his right arm and left leg were inflated with pendulous folded tissue (even while his left arm seemed utterly familiar). His body was strikingly asymmetrical, resulting in uneven movements. Eventually, Merrick reached the care of physician Frederick Treves and was welcomed in London's elite society. But such care was deliberately protective. Treves described how, earlier, Merrick "had been ill-treated and reviled and bespattered with the mud of Disdain."[11] Even under Treves's care, he went hooded and cloaked when traveling in public lest he spark incident. Merrick himself never stopped dreaming of being ordinary. Merrick's unusual form did not evoke fascination, but an alienation to be overcome.

FIGURE 17.2 *Joseph Merrick captured in an exceptional historic photo in 1889. Reviled in popular society as an abnormal "monster," he was celebrated in fashionable society as a learned gentleman.*

Consider also the case of Alexina/Abel Barbin.[12] Barbin was an intersex. Raised as a girl, s/he was "discovered" as male in 1861 and reassigned an "appropriate" identity. Scandal ensued. Now a male, he was denied marriage to the woman he had loved earlier as a female. He had to relocate. Seven years later, unable to readjust, the tortured Alexina (as she saw herself) took her own life. At that time, there was no place for those who did not fit "nature's" categories of male and female (essay 16). An intersex "monster" did not receive special regard. Rather, like Merrick (and unlike hirsute Petrus Gonsalus, profiled in essay 1), s/he was an outcast.

Ironically, then, the growth of science contributed to the displacement of monsters from their valued status. With explanations, the extraordinary became ordinary. With laws and patterns characterizing how nature was "supposed" to be, monsters became abnormal deviations. In answering the call of wonder, science ultimately destroyed that very sense of wonder.

Naturalizing the "Normal"

The concept of laws of nature has a powerful hold on our minds. The very language is highly charged. In human society, laws specify what we *ought* to do. They ensure social order. We tend to interpret laws of nature in the same way, as the guarantors of the natural order, profiling how nature *should* act. Once established, *descriptive* laws take on a *prescriptive* character. The laws of "normal" development easily become standards for how organisms "ought" to grow. The normal becomes *naturalized*, or apparently constitutive of nature's order.[13] At the same time, the abnormal comes to reflect undesirable disorder or chaos. Facts thereby become imperceptibly—but inappropriately—imbued with values. The irony of monsters is that while they are plainly products of nature, they are often viewed as "unnatural" because they seem to "violate" its "laws." The term "monstrous" now implies impropriety, not merely unusualness.

The effect of naturalizing the "normal" is not unlike a paradox of democracy. When one honors exclusively the wishes of the majority, the minority can be wholly disenfranchised. Such "tyranny of the majority" eclipses the political question of how to address dissent. In a similar way, undue focus on the laws of nature, or the normal, can eclipse understanding of exceptions or phenomena not fully described by the laws. One may call it, by comparison, the "tyranny of normality." Scientifically, it means our interpretations of nature may be skewed or incomplete. Culturally, it means monsters—according to the "natural" categories established by "science"—are shunned (or pitied) as abnormal, not welcomed or celebrated as unique.

The history of monsters, from marvel to pathology, invites us to reflect more generally on the tyranny of normality in biology and in our views of science generally. In what ways do we bias our image of nature by focusing primarily on "laws" of nature as fundamental or on the most frequently found form as a targeted norm? Sickle cell anemia, diabetes, autism, and invasions of nonnative species are no less expressions of biological processes than their often-privileged "normal" counterparts. In biology, absolute laws are rare. Exceptions abound. By establishing a sense of balance, we might foster a fuller, more complete appreciation of nature in all its diverse behavior.

The popular animated films *Monsters, Inc.* and *Monsters University* playfully upend conventional perspectives. The cartoon monsters are normal; the adorable child is an alarming "other." The humor conveys a lesson that the ordinary is a matter of perspective. So too, then, for the extraordinary. The challenge may be finding the appropriate perspective. Can we reimagine the seventeenth-century view of monsters as wonders, not freaks, without abandoning today's science? Might that perspective also, at the same time, help us view ordinary, or normal, organisms as wonders too?

18

To Be Human

Who are we? The question of human nature seems to haunt all disciplines. That may tell us how very "human" the question is.

Answers vary widely. Yet scientists—anthropologists, geneticists, ethologists, and developmental and evolutionary biologists—rely on observations and empirical data. Their conclusions thus seem more objective.

Biologically, humans are primates. Linnaeus perceived that, even before Darwin. We share our anatomies and physiologies with apes and chimps. But Darwin gave this relationship special meaning. He transformed abstract taxonomy into material genealogy. Ever since, we have characterized our species by its ancestry. Identity and history have merged. "Who we are" is now also the story of human origins: where we came from, how, and why.

Each new finding in human evolution seems to fascinate us. The sequencing of the human and chimpanzee genomes was big news, appearing on the cover of *Time* magazine.[1] Then came the Neanderthal genome.[2] "Ardi" (*Ardipithecus ramidus*) created a public sensation by replacing Lucy as the earliest known complete hominid skeleton, displayed dramatically on the cover of *Science*.[3] Then the human-like ape *Australopithecus sediba* sparked new controversy.[4]

Add to this buzz new exhibit halls on human origins at both the American Museum of Natural History and the Smithsonian.[5] And a cascade of books, whose topics range from surveying new fossils, vestigial traits, and genomes to profiling the uniqueness of our brains, bones, genome, and behavior.[6] And television specials.[7] We always seems eager for new perspectives.

But perhaps it is time to reassess this sacred bovine: that each new finding yields more-complete understanding of human nature. We might well reflect on our *past* efforts—with their notable errors and flawed assumptions. What might we learn from those missteps instead?

Seeking Uniqueness

Benjamin Franklin was reportedly among the first to celebrate humans as the only toolmaking animal. Later, evolution seemed to make sense of that. Our hands—especially with their opposable thumbs—once used for climbing trees, seem to have found a new adaptive function: to grasp tools, to shape them, to modify the environment and so enhance survival. Tool use also fit with the distinctive trait of walking upright: bipedalism apparently freed the hands to do their important work.[8] By the early 1960s, the uniqueness of humans as tool users was conceptually well entrenched.[9] Thus, when Louis and Mary Leakey identified the first fossil found alongside tools in 1960, they gave it landmark status as the first of our genus, naming it *Homo habilis*, or "handy man."

Yet with newer discoveries, humans could not maintain their unique status as tool users (Figure 18.1). Egyptian vultures use rocks to crack open tough ostrich eggs; California sea otters, to break mussel shells. The Galápagos "woodpecker" finch uses cactus spines to probe holes and collect ants. New Caledonia crows use twigs in a similar way. Even *Ammophila* wasps can use small pebbles to pack earth around nest entrances. Tailorbirds use found fibers to sew leaf edges together into nests, while *Polyrhachis* ants make nests by using a secreted adhesive to fasten rolled leaves.[10] Human tool use was not so unique after all.

Still, humans, like the early *Homo habilis*, seemed the only animals to *make* tools. The Duke of Argyll, Darwin observed, claimed that "the fashioning of an

FIGURE 18.1 *Tool users?* a: *A woodpecker finch using a cactus spine to probe for insects.* b: *An Egyptian vulture throwing a rock at an egg.* c: *A sea otter with a rock is has used to help open a shell for eating.* d: *A tailorbird nest, woven with a natural fiber.*

implement for a special purpose is absolutely peculiar to man." "This forms," he added, "an immeasurable gulf between him and the brutes."[11] Renowned evolutionist Theodosius Dobzhansky, too, noted that tool *using* may be instinctual, but "tool-making is a performance on a psychologically higher level."[12] And so human uniqueness was redefined: from tool use to toolmaking.

Yet we have discovered since that other animals, notably our primate cousins, do indeed make tools. Chimps crush leaves to make sponges to collect water from hollows in logs. They strip leaves from branches to use as probes for insects. They sharpen branches with their teeth for hunting and spearing bushbabies. They arrange two stones as "hammer" and "anvil" to open very tough panda nuts. Sometimes, they even use a third, wedge stone to level the pounding surface. What's more, chimps sometimes combine tools in complex sequences. In Gabon, they use a "tool set" of five tools to obtain honey: to pound, to perforate, to enlarge the hole, to collect, and to swab. Various chimp groups leave behind complete "toolkits," generally of about twenty tool types, distinctive of each group's culture. Primatologists now comfortably discuss "chimpanzee technology."[13] Well, so much for the uniqueness of *making* tools.

The unique trait then retreated to *teaching* tool use. Until, that is, adult chimps were observed to help younger chimps learn how to use the hammer-anvil technique (Figure 18.2). The chimps not only conspicuously demonstrated the method, but also sometimes corrected the orientation of the learner's stone hammer.[14]

Did concession ensue? Hardly. An early researcher of mountain gorillas contended, "there still appears to be a wide mental gap between preparing a simple twig for immediate use and shaping a stone for a particular purpose a day or two hence."[15] Only humans *plan* tool use. Or so it seemed at the time. Bonobos and orangutans have now demonstrated in tests that they can select appropriate tools, save them, and retrieve them for later use.[16]

Well then, as comedian Tim Allen has quipped, perhaps humans are the only animal to *borrow* tools?

The history of successive claims about tools exhibits an informative pattern. When observations falsify their hypothesis, scientists do not abandon the hypothesis. They change it. One unique trait simply segues to another. They have persistently sought to characterize humans as *unique*. Not just different in degree but qualitatively distinct, or unmatched.

This pattern also appears in the history of interpretations of language—another feature frequently profiled as uniquely human.[17] At first researchers noted that chimps, despite extended efforts, could not learn to speak. In the 1940s and '50s Keith and Catherine Hayes trained a chimp, Viki, to say "cup" and three other words. (Note that the awkward efforts were filmed with Viki dressed in a nice girl's smock, betraying the researchers' anthropocentric standards.) Further efforts failed: proof, apparently, that chimps could not manage speech or language. Or at least *human* speech. However, in the late 1960s Allen and Beatrix Gardner

FIGURE 18.2 *Tool teacher? An adult chimpanzee showing a juvenile how to crack open a nut using the two-rock "hammer and anvil" method.*

successfully taught another chimp, Washoe, about 250 signs of American Sign Language (ASL). The faulty assumption of the earlier work became painfully obvious. Chimps do not have the appropriate vocal apparatus for English language sounds. This does not mean they cannot communicate effectively or understand language. Why had no one noticed the conflation of language with human speech at the outset?

One can easily imagine the sequel: "Yes, but chimps don't. . . ." *Teach* language? Later, Washoe taught signs to her adopted son, Loulis. Or: chimps do not use language *on their own*? Eventually, other chimps who learned ASL were observed signing to one another without humans present. Boundaries of uniqueness are always being pushed back. And scientists always seem to finesse the uniqueness to something new.

So focus shifted elsewhere: to language structure. "The meanings of our sentences are composed from the meanings of the constituent parts (e.g., the words). This is obvious to us," noted one linguist. He then continued, ironically, "but no other animal communication system (with honeybees as an odd but distracting exception) puts messages together in this way."[18] Why dismiss the exception peremptorily as "odd" and "distracting" rather than pursue it as potentially significant?

Others retreated to defending the specific roles of symbols or of grammar.[19] Researchers then dutifully documented that chimps can interpret abstract lexigrams on keyboards and arrange them in meaningful sequences. And so on.[20] One need not subscribe to some Doctor Doolittle–type fantasy to acknowledge that animals (and even plants and slime molds) have some extraordinary systems of communication. Nor does one need any sophisticated science to know that humans communicate with remarkable complexity, significantly shaping their collective behavior. The question is why anyone feels a need to characterize human language as *unique*, rather than as an evolutionary derived variant.

One finds similar prejudices in other claims about human uniqueness. About the nature of play.[21] Or laughter.[22] Or art.[23] Or empathy and perspective taking.[24] Even the assumption that humans alone are moral creatures[25] now seems untenable. Macaques, capuchin monkeys, and vampire bats seem to express moral sentiments, reason about fairness, or act against cheaters (see essay 10).[26]

I have erred, too. Yes, I once professed that dental hygiene was uniquely human. We brush our teeth. Oh, and we floss, too. What other species advertises competing toothpaste brands? Yet one student chastised me: what about the symbiosis between cleaner wrasses and groupers (Figure 18.3)? Sadly humbled, I now concede that chimps use toothbrushes. They *make* toothbrushes, by chewing the ends of sticks. However, I continue to wonder about the uniqueness of plastic-coated teeth. And dentistry.

Despite the successive "yes, but" failures, scientists continue unabated the grand effort to articulate what makes humans unique.[27] Not long ago dozens convened to explore "human uniqueness and behavioral modernity" with renewed interdisciplinary vigor.[28] Alternatively, we might conclude, more humbly, that the quest for uniqueness itself seems to differentiate humans.

FIGURE 18.3 *The cleaner wrasse, which provides "dental care" to other fish by consuming parasites in their teeth and elsewhere.*

Seeking Distinction

Further lessons are found in the selective nature of the traits typically claimed as unique. Blushing is usually not high on the list. Darwin, however, called it "the most peculiar and the most human of all expressions."[29] Perhaps Darwin was just preoccupied with his social environment—the Victorian upper class, steeped with propriety. Yet Mark Twain agreed: "Man is the only animal that blushes." "Or needs to," he added.[30] And therein lies a clue. No one *wants* to be noted as unique for shameful or embarrassing behavior.

The exhibit on human origins at the American Museum of Natural History in New York is typical. Adjacent to the stunning models, fossil replicas, and human family tree, one finds a display on "What Makes Us Human." The answer? Intelligence, creativity, language, symbolic representation, music, art, tools: "the world of human expression." The key concept, according to the educator's guide, is, "Only modern humans create complex culture."[31] Note the selective pattern? The distinctions all mark things we value. Humans thus seem inherently privileged. Granted, the exhibit does credit bowerbirds for their "individual expression" and whales and birds for the "structure" in their songs.—But not without underscoring the limits of those abilities and reaffirming human superiority.

So, too, for the Smithsonian's exhibit on human origins. Except that it emphasizes instead social life and a reminder that "humans change the world."[32] Again, traits to bolster self-esteem. Why?

A more balanced appraisal might consider other human traits as unique: deceit, theft, murder, armed conflict, disparity in social levels, enslavement. Although one would have to reject those traits, too, since they are all found in other animals as well. We could revise the list of traits to: weapons of mass destruction, Ponzi schemes, high-frequency obesity, and large-scale devastation of ecosystems. Those just might pass the test of uniqueness. But would anyone entertain them as defining human nature?

Alternatively, one might consider less impressive but still unique traits.[33] Such as the uneven distribution of body hair. Or tears.[34] Or prominent chins, a distinct result of evolving cranial development. Or the capacity to choke on food.[35] Or back pain and knee injuries, to go along with our walking upright. Or genetically based susceptibility to tuberculosis and malaria.[36] Along with the ability to be rational is the ability to rationalize. As much as we can laugh and tell jokes,[37] we can also tell *bad* jokes. We may not use language exclusively, but we are alone in texting messages while driving motor vehicles. Why do profiles of human identity not highlight these more prominently? Ultimately, human uniqueness is not just about difference. It also implies value.

For an interesting glimpse of general opinion on "what it means to be human," one may visit the public comment website for the Smithsonian's exhibit on human origins (http://humanorigins.si.edu./about/involvement/being-human). A sample:

to wonder
to laugh and to cry

to regret and to expect
to imagine I'm a horse with my friends (age 4)
to be capable of the best and the worst
to honor your ancestors and your children
to know more than you should and think less than you could
to seek out the real magic of the world around us.

These responses illustrate how the question is typically interpreted. Not descriptively, but normatively. The meaning is based on what people think humans *should*
be. Or what they want themselves to be. Not what we are. While the phrase "human
nature" appears objectively neutral, too often it is permeated with ideals or ideology.

Scientists are no exception, despite the rhetoric about their objectivity. The history of science on human origins is littered with normative judgments, as profiled
above. Indeed, one can often infer a scientist's *personal values* from his or her *scientific claims*. For example, the common emphasis on intelligence, cognition, rationality, and culture? Endorsed by (surprise!) scholars. For anthropologist Ian Tattersall,
ideology is visible in his unguarded praise of technology and of the "restless innovative spirit."[38] For neuroscientist Michael Gazzaniga, it is betrayed in comments
about tools and Maserati cars or bipedalism and Italian designer shoes.[39] For him,
human uniqueness seems determined by the criteria for a good date: conviviality
(social networking, consciousness, and empathy), "intelligent" conversation, appreciation of the arts, and trust, or a "moral compass" (he addresses each in a separate
chapter). One would be well advised to take even "scientific" claims about humanness with a grain of salt—or at least with a heavy dose of critical skepticism. That
may include the very label *Homo sapiens*, or "wise man."

Naturalizing Values

Historically, then, the science of human nature has proved treacherous. In search of
uniqueness, it has shown systematic overstatement. It has discounted similarities as
well as differences of degree in securing a special status for humans. By focusing on
certain traits, it has excluded others, fostering a misleading portrait that privileges
our species. But none of this should surprise us. When values are at stake, science is
susceptible to the *naturalizing error*.[40]

Oddly, the features that scientists typically use to characterize humans do not
require science at all. No systematic observation. No measurement. No sophisticated training is needed to see that humans dominate the planet with cultivated
land, highways, oil rigs, coal mines, and depleted fisheries; that we form monumental social networks and communicate globally with cell phones and satellite
TV; that we create governments, prisons, peace prizes, and disaster relief efforts;
and so forth. Why the science, then? The historical lapses show vividly how *the science functions primarily at a rhetorical level: to "justify" the value-laden claims about
human uniqueness and distinction*.

Science here may seem to simply render the facts of nature—or of human nature.
But these facts, or images, of nature are not neutral or balanced. They never were.

Even if "true" at some level—supported by observations or other evidence—they are highly selective. Identifiable values guided the science. Worse, little vestige or trace of the selectivity remains. Nature is visible; the values are not. Through science, the values have become *naturalized*, and henceforth masquerade as undeniable facts.

The appeal to nature is especially well illustrated in the popular fascination with the genetic basis of evolutionary changes. No one needed a genome project to conclude that humans are closely related to chimpanzees and other primates. But genes are now paraded as the essential, foundational cause of the similarities (essay 19). That's why the comparison between human and chimp DNA makes the cover of *Time*.

Yet we finesse the new genetic findings as much as any other evidence. Our genetic correspondence with chimps is a remarkable 94–99.2%, depending on how one measures it—98.7%, if based on nucleotide sequences.[41] But our response has not been to embrace the similarity. Rather, we typically underscore the remaining difference and its apparent importance.[42] We simply wonder even more deeply how the dramatic—and obvious—differences can be encoded in so few genes. That interpretation conveniently bypasses the more complex and open-ended behavioral or cultural dimensions.

No one questions nature. Anything declared natural seems an inherent fixture of the world, transcending human culture and values. Unchangeable, hence unchallengeable. The "nature" in "human nature" thus functions more as justification than description. Cognitively, our minds all too readily imagine that our personal values express some universal "human nature." When we find evidence that aligns with our values, our critical faculties go on holiday. Accordingly, we cannot regard the role of science in naturalizing human values as mere happenstance, even if subconscious. One may well doubt that a respectable science of human nature is ever possible.

Being Human

In his great system of classification, Linnaeus did not describe humans using standard anatomical structures. Rather, he left readers the Socratic injunction "Know thyself." History may now indicate that this approach is risky. Human values may easily eclipse a fully informed view. Science can naturalize, and thereby rationalize, prejudices. (If we could only secure an analysis from another species!)

What, then, is the remedy? Perhaps our enduring fascination with human origins and evolution offers a first step. Perhaps we can emphasize not our uniqueness, but our continuity with other animals.[43] In his bestseller *Your Inner Fish*, Neil Shubin nicely portrays how the human body itself embodies its history, with links to fish, worms, fruit flies, and jellyfish as well as primates.[44] He models how one may convey connectedness without falling prey to the other, equally inappropriate extreme, the view that we are "nothing but" animals.

The paleontological lessons are echoed by discoveries of the Human Genome Project. Similarities at the molecular level remind us of our position in the evolutionary tree. That is, we share genes as well as body structures with other organisms, as exhibited by such genes as *Noggin, Sonic Hedgehog, Hox*, and *PAX2*. The genome gives evidence of our common ancestry—our extended genealogy—without diminishing our own humanness in any way.

Most important, perhaps, we can reflect about why we seem to care so deeply about the very question "What does it mean to be human?" And what we expect of a scientific answer. And why. That reflection may itself reveal part of the answer.

19

Genes R Us

DNA fingerprints are not prints of fingers. So why the name? The "fingerprint" label, of course, conveys far more than some pattern of swirls, whorls, and arches on the skin. As celebrated in detective lore, fingerprints are emblems of uniqueness. DNA can thus form a "fingerprint" by establishing personal identity. Genes are often characterized as "information." Thus, DNA "codes for" an organism's unique traits. In terms of uniqueness and developmental causality, then, genes seem to underlie human identity. Yet with deeper reflection, one might find this commonplace association spurious and misleading: another sacred bovine.

Ironically, perhaps, DNA fingerprinting reveals very little about an individual's DNA, or genome. The technique does not exhaustively profile every form of every gene, as many imagine. Nor does it even *sequence* the DNA. Rather, it focuses on a rather incidental feature of chromosome structure: differences in noncoding sections of DNA. There, short "nonsense" segments are repeated. The number of repeats, however, varies widely among individuals. Thus, they are convenient markers, or *indicators*, for identifying a particular organism—or a potential criminal suspect. Each person's DNA may well be unique, but only a small and physiologically insignificant fragment of it is needed to *identify* a particular individual.

Other biological features function as identifiers, as well. Forensic scientists have long relied on fingerprints and "mug shots," both introduced into criminology by Charles Darwin's cousin Francis Galton. They also use hair, skin tone, blood and tissue type, and voice sonograms. Some high-tech security systems—including ones adopted by US immigration—use eye scans. These record the unique pattern of the eye's iris. (Blood vessel patterns on the retina work as well.) In all these cases, the aim is unambiguous identification. What matters is diagnostically unique properties. So these particular features are effective *indicators*. At the same time, their functional role is trivial. They are biologically insignificant. They hardly profile someone's sense of self. Nor do they fully characterize who they are (personally, culturally, or even biologically). *Identification* and *identity* are distinct. A unique feature is not necessarily an important feature.

Even genetic uniqueness falls short of defining identity. Consider, for example, identical twins (Figure 19.1). Identical twins share a genome. They fail the test of genetic uniqueness. Yet regarding them as distinct individuals is not problematic.

(All too aware of their individuality, one is often challenged instead *to* carefully discern which twin is which) Each twin has his or her own identity. They have separate names, independent lives, recognizable personalities—all quite apart from their similarity in appearance. Genes, in these cases, hardly establish identity.

Consider conjoined twins, too. The famous Hensel twins, Abigail and Brittany, share not just genes but the same body.[1] How does one characterize *their* separate identities as unique genetically or biologically?

Genetic uniqueness nevertheless seems central when considering the growing prospect of human cloning. In popular images—whether horrific or humorous—cloning recreates the original individual. Some envision armies of anonymous clones. Individuality would be lost in an ocean of biological Xeroxes. Others imagine that a genetic clone would somehow duplicate memories, too. Could one distinguish copy from original? Cloning thus seems to threaten unique identity. Some people have responded by declaring copyright on their genes. The law, they imply, will acknowledge that one "owns" one's genes.

Yet all such fretting seems blind to the lessons of identical twins. Twins are clones. Can one violate the copyright of the other? Are their memories identical? Are they empty, mindless drones? Of course not. Twins are individuals, as any clone would be. Clones are like twins displaced in time. That might challenge our conventions, but not our notions of identity. Confusing genes with identity may lead one wildly astray.

The tendency to locate unique identity in genetics extends to the species level, as well. Each species has its own genome. The differences between genomes are clearly linked to the differences between species. Variations in homologous genes,

FIGURE 19.1 *Clones, or just twins with independent identities?*

for example, are effective tools for showing relatedness. From them, one may construct evolutionary trees. Diverging gene sequences reflect diverging lineages. Genes thereby map common ancestry. They help mark our species' genealogical identity. They reveal our kin and history.

Even more, the degree of genetic difference seems to indicate the degree of relatedness. Closely related species share more genes. But note that the focus is just on *differences*. One may wonder here, as in the case of fingerprints, just what such differences represent. For example, the variations used to map evolutionary relationships typically do not document functional differences. The amino acid sequences of the proteins cytochrome c and hemoglobin may vary slightly in different organisms, but they fill the same physiological roles.

Not all genetic differences are significant, even if valuable in species-level identification. Consider a prospective "human" gene for how hair is distributed over the body surface. It certainly would be biologically unique among primates. Does it matter? Perhaps we should rethink patchy hair as important to our identity? Or perhaps not. Similar reasoning might apply to genes that lead to subtle differences in immunology or sensitivities in diet. It is not the genetic differences alone that matter to human identity. Rather, it is how we regard the organismal trait associated with the gene(s). Someday we may well find a gene responsible for differential growth of the cranium and thus brain size. But such a genetic difference will seem significant only given some vague notion of intelligence. Our conceptions of identity tell us *which* genetic differences seem to matter, independently of the genes themselves. And natural selection acts on the phenotype, not genes. Traits, not genes, establish meaningful identity.

Genes nonetheless seem foundational biologically. Thus, many persons tend to regard a change in the genome as an *essential* change in the species. New species evolve when the *genes* change. Each species' set of genes seems a naturally defined and inviolable identity. For example, hybrids are typically cast as "monsters," rather than playful inventions or fruits of nature's creative powers (essays 1 and 17). Concerns about genetically modified organisms, or GMOs, likewise rarely focus on the modification itself. What seems to matter is disturbing the species' genetic identity. Even modification of one gene alone can apparently disrupt a species' integrity. Transplanted genes are not viewed like transplanted organs. Genes seem to signal that the nature of the species is at stake. Ironically, humans have been altering organisms genetically throughout history. Domesticated plants and animals are all "genetically engineered" from wild species. Common bread wheat (*Triticum aestivum*) is even a hybrid of *three* species. So bread, an iconic staple, has long been made from a genetically modified organism. Yet juggling of genes may be easily overlooked when one focuses on cross-breeding at the organismal level. One may fail to see the "genetic modification" across species. Common concerns about GMOs underscore the depth of beliefs about genes as fundamental to identity (essay 27).

Belief in genetic uniqueness as group identity underlies most racist thinking, as well. A claim about racial difference is not just about biological variation. It is about *essential* differences. Such differences seem "essential" merely because they are genetic. "Genetic" implies "natural," or fixed in the world's organization. Race thereby can seem independent of human culture or human interpretation. By contrast, racial categories are much harder to rationalize as given by nature when one views identity as equally shaped by a complex context and by an organism's social and ethnic environment. Highlighting genetic differences at the group level also eclipses thinking about differences within groups. Unwarranted stereotypes easily substitute for individual features and variation—further masking the creative role of culture. Conceiving racial identity as genetic may be all too convenient to some, but it is biologically unjustified.

Genes ultimately seem closely related to identity because of their generative, or developmental, role. Genes seem singularly important causally—in guiding individual, species-specific development and determining all cell functions mediated by proteins or enzymes. Genes seem the root cause of every biologically important detail. Why should they *not* be viewed as central to identity?

Genes are indeed part of the causal story. But focusing on their differences (again) misrepresents their relative importance. In a widely used metaphor, genes encode information, like MP3 files or CDs or Braille. The metaphor can help show how genes are causally limited. MP3 files need MP3 players; CDs, CD players. An instruction in Braille is meaningless without the ability to read Braille. Just so with genes. DNA needs ribosomes. (One might note, too, the role of other cell structures and processes, including various enzymes that transcribe the DNA, edit the messenger RNAs, and ferry them across the nuclear membrane.) CD or MP3 players are also idle without energy. Likewise, protein synthesis requires energy in the form of ATP and GTP. In a cell without energy or ribosomes, genes are idle. They do not "express" themselves. A heap of DNA, by itself, is causally inert. Organisms reproduce through whole cells, not genes alone. The cellular context, with all its material "machinery," is equally inherited. Genetic "information" may be significant in guiding alternative physiologies, but the genes themselves do not fuel the process. Focusing on genes simply because they differ from one organism to another distorts the whole causal picture.

Just as the genes rely on the cell's internal environment, so too does the developing organism rely on many nongenetic factors. These may include (at least) temperature, pH, available nutrients and molecular building blocks, cell-cell contact, hormones, and other chemical triggers. Perhaps light or gravity, as well. The environmental elements provide the developing organism with "information" of a different type. Cells that share genetic information (like identical twins) thus do not all develop alike. Change the environment and a different cell develops— or none at all. A parent that provides the appropriate environment to an egg helps ensure that these critical causal factors are present. The environment can be inherited, too. To the degree that the environment seems predictable from one generation to the next, its causal role may seem invisible. But it is no less important in guiding an organism's development. Genes and environment together shape identity developmentally.

From cloning to GMOs to racism to DNA fingerprint identification, the reduction of identity to genes is problematic. More than genes are involved. For example, a sense of our species' identity might well include a potential for humor, morality, or appreciation of music. Yet we cannot trace these directly to discrete genes or genetic differences from our closest relatives. No doubt there is a "genetic basis" for the potential of brains to express these traits. But this hardly informs us how the behavior develops. Genes might well contribute to a concept of identity. But they cannot substitute for a phenomenon that is ultimately more complex and established in a cultural context.

Addressing how genes and identity relate may seem a subtlety of interest only in leisurely academic reflection. But the notion has strong political overtones. If one believes that humans are determined primarily by their genes, then biology, not society, is responsible for any differences in the human condition. Variations in social status, for example, seem to arise from "natural" causes. Social inequities can easily be attributed to biological stratification, not human politics. The status quo will seem inevitable, and unchangeable. Such a framework, of course, tends to justify current distributions of power and privilege. Worse, a political position—a value—might seem derived from fact. Genetic reductionism seems scientific, but like other forms of biological determinism, it is political ideology. For this reason alone, one might be concerned about what is said about genes, whether inadvertently or with a sense of purpose.

20

The Peppered Moths, A Study in Black and White

Organisms seem exquisitely adapted—like the peppered moths (*Biston betularia*), camouflaged to avoid predation (Figure 20.1). The story behind their cryptic coloration and pattern is easy to appreciate when looking at both black and mottled white moths set against contrasting backgrounds: one black, one mottled white, like the moths themselves. What half-witted bird would not prey on the obvious moth? This would clearly change the genetic make-up of a population. This popular pair of photos explains natural selection in an instant. The images are so widely reproduced, in biology textbooks and elsewhere, that everyone seems to know the case of the peppered moth.

The case gained renown through Bernard Kettlewell. In the 1950s he investigated the survival rates of the moths in the contrasting forests of Birmingham and Dorset (ostensibly portrayed in Figure 20.1). But Kettlewell's landmark publication, *The Evolution of Melanism*, also included another image (Figure 20.2).[1]

FIGURE 20.1 *Peppered moths (*Biston betularia*) staged on model backgrounds of lichen- and soot-covered tree trunks, illustrating the adaptive significance of cryptic coloration.*

On the top right, it displays the two familiar forms of the moth: *typica*, the once-common "peppered" form (no. 2), and *carbonaria*, the nearly black form that proliferated later (no. 1). Arrayed on the left, however, are five other specimens of the *same* species, each exhibiting an intermediate darkness. Together, they constitute a third form, known as *insularia*. That is, a series of relatively unknown light and dark forms fills the gap between the two well-known extremes. Do the *insularia* moths matter? How might simplifying the wide range of forms to just two shape an image of natural selection? Or of nature, generally? In what ways does the difference between the simple story and a more complex reality ultimately affect our thinking? Perhaps we should challenge the implicit assumption (yet another sacred bovine) that nature, and science too, can effectively be reduced to black and white.

FIGURE 20.2 *The untold story: some peppered moths are intermediate in coloration—the* insularia *forms on the left.*

Reducing Nature to Black and White

One easily finds specimens of *insularia* in museum collections, Kettlewell noted. And he included them in his field studies. Having recruited observers from around Britain, Kettlewell catalogued the relative frequencies of all three forms in various locations. The incidence of *insularia* was sometimes 40% or more.[2] *Insularia* was no trivial exception.

Still, while Kettlewell documented *insularia* in his research, it became eclipsed in subsequent renditions of his work. For example, the cover of his 1973 book sported the now-canonical images that, like Figure 20.1, omit *insularia*. Kettlewell himself seemed to promote the streamlined story publicly. In his widely reprinted 1959 *Scientific American* article, one figure displayed the relative percentages of *insularia* at various locations, while the fuller description in the main text, ironically, failed to mention them at all.[3] By hiding the real complexity, the now-familiar presentation of the peppered moth reduces nature to black and white.

Of course, one should not thereby conclude that Kettlewell's studies fail to demonstrate the role of natural selection in evolution. However, the simplified image can significantly shape how nonbiologists think about natural selection. There seem to be, after all, only two types of moths. And only two types of environments. The struggle for existence seems to pit just two types against each other: the fit versus the unfit. Competition seems to have only winners and losers. Either-or. Win-lose. Moreover, the environments are portrayed as unidimensional and uniform. A well-informed view of selection involves *differential* survival and *differential* reproduction in sometimes *heterogeneous* environments. But the potential power of the more sophisticated view is eclipsed when nature is cast in black-and-white terms.

For example, the reduction to black and white affects metaphors about nature, and hence perceptions about what is deemed "natural" in human culture. Many people already tend to conceive natural selection in stark terms (essay 8). Survival is life or death. The black-and-white story conveys how competition occurs, whether for peppered moths or for humans on the football field, in business, or in congressional debate. The simplified version of the peppered moth story is thus far from idle. When simplicity is projected onto nature, it seems to exist independently of human interpretations. It seems inevitable. The black-and-white rendition of the moths helps *naturalize* simplicity falsely into the fabric of the world.

One may surely profit from simplifying concepts for convenience and ease of understanding. But the qualified proviso of simplification can easily be forgotten. The artificial simplicity thereby becomes standard. It becomes normal. In addition, when a case such as the peppered moth is paraded as a prime example, it becomes emblematic of how nature *is*. One loses any trace of the original as actually being more complex. Worse, in educational and many public media settings, such abstractions are repeated again and again—establishing a pattern. The repetition nurtures a habit of expecting simplicity. Simplicity becomes a significant problem when it supplants and replaces the more complex reality.

The misleading imagery of the moths is compounded when the discussion turns to genetics. With just two forms of moth—dark and light—one easily assumes that there is just one underlying gene, with one "black" and one "peppered" alternative.

That is the traditional Mendelian model, as popularized too for a public audience by Kettlewell.[4] This assumption also fits the most-basic models of population genetics, frequently taught in biology classes using the peppered moth case as an example. Unfortunately, the spectrum of *insularia* moths immediately implies greater complexity. In his scientific publications, Kettlewell imagined multiple alleles, or several alternatives, for a particular gene.[5] Other experts now propose the overlapping effect of many genes. Most importantly, one "on/off" genetic switch is insufficient to explain what we observe. One false interpretation of simplicity (in coloration) fosters another false interpretation of simplicity (in genetics). Indeed, the case is only one of many that help perpetuate the mistaken notion that Mendelian dominance itself is fundamental (essay 22).

Ironically, the portrayal of simplified genetics for the moths tends to reinforce the original impression that evolution is also simple. Everything about the peppered moth case is black and white, not just the moths. The agreement among apparently different perspectives further helps erase awareness of the transformation from a once-complex reality. The crafted illusion of simplicity in nature is complete. The black-and-white story misleads us about nature, while naturalizing simplicity in nature itself.

Reducing Science to Black and White

Nature is not the only feature of the peppered moth case that is simplified. So, too, is the process of science. Kettlewell's field studies are now widely celebrated. Indeed, many biology textbooks honor them by describing the experiments and noting their elegance and significance. Some even include tables or graphs of the original published data. However, these "textbook histories" can be greatly streamlined, too. Like the image of the moths themselves, they may shutter out important information—with profound effect.

Kettlewell conducted many field studies. He measured the different survival rates of the moths in different forests where the tree surface background varied. He raised moths, marked them, released them into the wild, and then counted those that could be recaptured. In supplemental work, he conducted tests and made other field observations to ensure that the counts were accurate. Kettlewell boldly presented his results as "Darwin's missing evidence."[6]

But the story is typically tidied. For example, many textbooks highlight Kettlewell's exemplary scientific practice in using two contrasting environments to show their effect on natural selection. In the dark, polluted woods near Birmingham, the melanistic forms (*carbonaria*) were recaptured twice as frequently. On the other hand, in the lichen-covered woods of rural Dorset, the speckled forms (*typica*) were twice as likely to survive. The coupled investigations exemplify a colossal controlled experiment, with the environment as the sole variable. That's the familiar textbook story.

As historian Joel Hagen has observed, however, originally (in 1955) Kettlewell presented only data from Birmingham.[7] He made no reference whatsoever to Dorset. Nor did he give any hint that his study was incomplete or preliminary, or that readers could expect forthcoming complementary data. Why? If the Dorset data were so crucial, was the first study flawed? What does Kettlewell's apparent omission mean?

Hagen considers several reasons why Kettlewell might have published the Birmingham results alone. Originally, perhaps, Kettlewell did not see the control as important. This seems likely, given subsequent criticism of his work, his personal correspondence, and the apparent timing of his plans to add Dorset the following year. In that case, the need to address criticism would have motivated Kettlewell's extended study, not an initial perception of the need for clarity. That is, he did not conceive the entire set of experiments in a single flash of insight. Rather, he patched together two separate studies. This scenario differs from the stereotypical image of great scientists working by way of "Eureka" moments. Scientific achievement can involve less extraordinary modes of thinking and working (see also essay 21).

In another scenario, Kettlewell could not afford to run both full-scale experiments simultaneously. The release-recapture method involves substantial labor. Kettlewell was working alone at first, without funds to hire a field assistant. Later, Kettlewell enlisted his wife and son to help. Travel between the two sites would also have been problematic. Or perhaps Kettlewell began with a pilot study, which yielded unusually favorable results. Or he might have rushed to publish simply to establish his priority. One of the hidden tasks in a recapture study is raising the hundreds of organisms for release—and having them all ready at the appropriate time of year. That would have meant breeding moths in cages and sorting each form—not as simple as one might imagine. Was Kettlewell limited by sheer logistics?

All these possible alternatives reflect the complexities of doing science—labor, cost, ambition, developing reliable technique, maintaining lab organisms, and responding to peer criticism. They show that the process of science is not so black and white as the conventional "scientific method" implies (see also essays 5 and 10). A story that celebrates science but excludes these features implicitly conveys that they are irrelevant. Yet they would have been relevant in interpreting Kettlewell's claims at the time. By portraying an exemplary case of science as simple, a short history implies that all science—or all *good* science—is simple.

Textbooks also tutor students to see Kettlewell's studies as well designed, definitive, and thus beyond all doubt. This earns them their classic status. Philosopher David Rudge paints a more complex picture.[8] He focuses on Kettlewell's central claims about bird predation. At the time, many scientists doubted whether birds preyed on peppered moths at all. So Kettlewell collected other data to ensure that his recapture rates reflected predation, not some other environmental factor. For example, he enlisted ethologist Niko Tinbergen to film birds eating the moths in the wild. He checked whether the traps tended to collect one form more than another. He monitored dispersal of moths away from the study area. Texts rarely discuss these tests, Rudge notes. To further isolate predation as the chief causal factor, Kettlewell would need to have tested the presence versus absence of birds, even if it required an unimaginably large enclosure to exclude bird predators. Alternative explanations seem not to have all been ruled out.

The comparison of results in the Birmingham and Dorset forests—as depicted in Figure 20.1— is often considered critical. But Rudge contends that the key variable was not polluted versus unpolluted woods. Testing the role of pollution would have required the two sites to be parallel in all other relevant respects. Were they?

Kettlewell recaptured nearly twice as many moths in Birmingham as in Dorset. Why? Did the release of a large number of moths alter predation rates? Can we safely generalize from the release-recapture moths to moths in nature? When viewed more closely, the path to secure conclusions is more complex. In the textbook image, science sorts things crisply into true and false, proven and unproven, without any "shades of gray." No partial or tentative conclusions. No residual uncertainties. Science, too, is often reduced to black and white—like the moths themselves.

Reducing Science in Society to Black and White

The simplifications of the cases of the peppered moths and of Kettlewell's science, of course, are not isolated—nor even atypical. Meselson and Stahl's work on DNA replication, once described as "the most beautiful experiment in biology," also has a hidden complexity behind the scenes (essay 4). The Keeling Curve, which critically documents the long-term rise of carbon dioxide in the atmosphere, also hides its complex history of patchwork funding (essay 2). Ultimately, a fuller understanding of such historical cases matters, because they form a context for interpreting current social issues informed by science.

Citizens or consumers who encounter exclusively simple accounts—especially when rendered as exemplary science—inevitably develop a view that nature and science *are* simple. They become conditioned to *expect* simplicity in scientific evidence and reasoning. That, in turn, shapes how they think about science in society.

But to be well informed about science means in part to understand that it can involve uncertainty, ambiguous results, and apparently inconsistent evidence. Studies may be incomplete. Data may be subject to contrasting interpretations. Different studies may support contradictory conclusions. The subject itself may be complicated, even if well understood. Consider the case of the fluoridation of public water supplies in the 1950s and '60s.[9] Originally, research indicated that modest fluoridation could help prevent tooth decay. Yet some people found the prospect of a chemical's being introduced into their water without their individual consent to be an imposition reminiscent of totalitarian states. So "antifluoridationists" arose. What might have unfolded as a discussion about the role of government in public health, however, was transformed into a debate about the science. That is, each political position appealed to a scientific characterization of nature as the ultimate benchmark. Advocates thus pointed to the benefits of dental health. Meanwhile, critics pointed to the risks of excess fluoride (fluorosis), possibly including cancer. The evidence became dichotomized into simple categories: "for" or "against" fluoridation. The supposed voice of nature, through science, was polarized.

At that point, the science might have informed a discussion of how to manage the "pros" without experiencing the "cons." But each side presented nature as speaking *exclusively and decisively* for one answer. That is, *both* sides adopted a posture that science provides a black-and-white view of the "nature" of fluoridation. And so debate festered.

With no clear resolution at the evidential level (nature being complex, as it often is), arguments shifted to the credibility of the scientific claims and of those who made those claims. The assumption was that there was either good science, defended by good scientists (who heeded the evidence), or bad science, promoted by bad scientists (who were biased by politics). And, of course, the judgments of bias were based not on any independent criteria but on whether the claim under the banner of science aligned with one's own original ideology. The nature of scientific integrity and expertise was also reduced to stark either-or, black-and-white terms.

The fluoridation controversy vividly illustrates the outcome of the simple views noted in the peppered moth case. Namely, the assumption that science is certain and unambiguous—the black-and-white model—impairs fruitful reflection and discourse on socioscientific issues. Nor should one regard the case of fluoridation as unusual. Science was cast in similar extremes in cases involving nuclear power, a 1986 state referendum on mandatory HIV testing, the cold-fusion fiasco of 1989, and the teaching of evolution.[10]

The science that informs our social and personal decision is complex more often than not. The safety of Vioxx as a pain reliever, the prediction of earthquakes, the reasons for the hole in the ozone layer, the causes and severity of climate change: these are not resolved with simple either-or results, like the black-and-white versions of the peppered moth case.

It may seem that simplification is essential. How else can science be communicated to the public or understood by nonexperts? But a simple view not coupled with an awareness that the view *is* simple can be misleading. One needs a general understanding of *how* simple views can be misleading, perhaps as illustrated vividly by just one case, such as Kettlewell and the peppered moths. A view of the complexity in science allows one to see the simplified accounts as inexact and incomplete. Simplicity itself is not misleading. Rather, the problem is enshrining it as a model of science or of nature. One must understand the tendency to naturalize simplicity. Elsewise, one may succumb to perceiving the world in black and white, where a mere wish for simplicity eclipses a more textured reality.[11]

PART VI

Myth-Conceptions

21

Alexander Fleming's "Eureka" Moment

Alexander Fleming shared the 1945 Nobel Prize in Medicine for the "discovery of penicillin and its curative effect in various infectious diseases." According to the story as commonly told, a stray spore of *Penicillium* mold, borne by fortune through an open window, landed on an open bacterial culture in Fleming's lab in 1928. Fleming later happened to notice a clear zone on the plate, where no bacteria grew (Figure 21.1). "Seeing that halo was Fleming's 'Eureka' moment, an instant of great personal insight and deductive reasoning," *Time* magazine celebrated on its website 100 Persons of the Century. "When it was finally recognized for what it was—the most efficacious life-saving drug in the world—penicillin would alter forever the treatment of bacterial infections. By the middle of the century, Fleming's discovery had spawned a huge pharmaceutical industry, churning out synthetic penicillins that would conquer some of mankind's most ancient scourges, including syphilis, gangrene, and tuberculosis."[1] Millions of lives were saved, it seems, based on one astute observation. Fleming's story reflects an ideal about science: a moment of genius cascades into monumentality.

At the same time, the discovery of penicillin is also one of the most famous cases of chance, or accident, in science. Fleming himself often underscored the role of chance in his work. Despite his numerous honors and awards, he was fond of reminding others: "I did not invent penicillin. Nature did that. I only discovered it by accident."[2] Yet one may recall Pasteur's famous dictum that "chance favors the prepared mind." That is, Fleming was surely being modest. He seems not just lucky. He seems exceptionally perceptive. What but intuitive genius could have guided the "great personal insight and deductive reasoning" hailed by *Time* and marked by the Nobel Prize? The role of "Eureka" moments in propelling scientific progress is certainly ingrained in popular lore—so deeply as to escape questioning.

However, with a bit more historical awareness, one may well wish to challenge this sacred bovine. Ironically, the episode exhibits even more "chance" than is often told. Many others had earlier noticed just what Fleming did. Why did their observations lead no further? What did Fleming actually do? With close inspection, his achievement seems to have depended more on details of context than on any special personal attributes. In addition, the traditional account obscures a considerable amount of raw labor and persistence in identifying penicillin's clinical efficacy.

FIGURE 21.1 *Alexander Fleming, in a familiar restaging of his "chance" discovery of penicillin. Originally he did not think the drug had significant therapeutic potential.*

Without several researchers who aggressively pursued the uncertain prospect in Fleming's initial observation, his "discovery" would likely have remained invisible to history. A fuller account suggests a less romanticized view of science. The discovery was due primarily to circumstance, or contingency of events, and to concrete work.[3] The conventional story thus allows us to ultimately reflect on how we tell—and retell—scientific stories.

The Context of "Chance"

Renewed interest in the history of Fleming's work began many years ago when a bacteriologist in London noted that the windows of Fleming's lab at St. Mary's Hospital could not open. How could a stray mold spore have floated in an open window, even by chance? Second, he observed, spores of *Penicillium* will not germinate under the conditions described by Fleming. Someone else then observed that the particular species of *Penicillium* involved in Fleming's "discovery" would not likely have been floating in the air of London. Though common bread mold is a variety of *Penicillium*, it was the much rarer *P. notatum* that produced Fleming's penicillin.

The most likely source of the mold, it now appears, was a mycology lab downstairs from Fleming. There were likely spores all over the building. In addition, Fleming was never known for neatness in his lab. Open cultures would not have been uncommon. It almost seems inevitable, then, that the mold would have contaminated one of his cultures sooner or later.

The conditions of contamination would also have been important. Fleming believed, based on his earlier discovery of the enzyme lysozyme, that penicillin acted by breaking open bacterial cells. That would certainly have accounted for the watery appearance of the culture where the bacteria were absent. In that case, the spore would merely have needed to land on the culture plate—just as Fleming reported it. But we have since learned that penicillin acts by blocking the synthesis of chemicals that bacteria use to *build* cell walls. Penicillin does not kill bacteria outright. Rather, it prevents their effective reproduction. A spore landing on an existing culture would thus have had little immediate observable effect. The mold would have had to establish itself first if it was to prevent the further growth of bacteria. Temperature conditions while Fleming was away from his lab on vacation may have allowed this, or Fleming may have inoculated a plate that was already moldy. In either case, a stray mold spore alone would not have created what Fleming said he observed.

The circumstance whereby Fleming noticed the original culture also seems quite improbable. Fleming did not notice the mold's effect while routinely examining his cultures, though he did inspect them when he returned from his one-month summer vacation in 1928. In fact, he had discarded the now-famous culture and left it to soak in a tray of Lysol. A former member of his lab stopped by to visit, however, and Fleming showed him several cultures. Among these, he casually selected the critical culture from the top of the discarded stack, where it had escaped the liquid disinfectant. Only then was Fleming struck by the unusual pattern of growth. In his initial paper Fleming wrote only, somewhat cryptically, that it "was noticed."[4] Not so heroic, perhaps.

Penicillin in Historical Context

Fleming was certainly not the first scientist to have noticed the antibacterial effects of molds.[5] In 1871, Joseph Lister (noted for introducing antiseptic practice into surgery) had found that a mold in a sample of urine seemed to inhibit bacterial growth. In 1875 John Tyndall reported to the Royal Society in London that a species of *Penicillium* had caused some of his bacteria to burst. In 1877 Louis Pasteur and Jules Joubert observed that airborne microorganisms could inhibit the growth of anthrax bacilli in previously sterilized urine.

Most dramatically, Ernest Duchesne had completed a doctoral dissertation in 1897 on the evolutionary competition among microorganisms, focusing on the interaction between *E. coli* and *Penicillium glaucum*. Duchesne reported how the mold had eliminated the bacteria in culture. He had also inoculated animals with both the mold and a lethal dose of typhoid bacilli, showing that the mold prevented the animals from contracting typhoid. He urged more research. But following his degree he completed army service and never returned to research. Chance, here, ironically worked against discovery.

Several other researchers (almost certainly unknown to Fleming) had noticed the effects of *Penicillium* molds on bacteria. Fleming was not unique (Figure 21.2).

But noticing a phenomenon does not always mean that one fully appreciates its potential meaning. The chance in Fleming's case may have been less the appearance

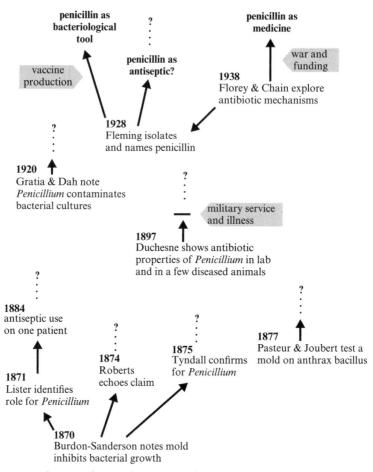

FIGURE 21.2 *Alternative historical trajectories that, with other contexts or "chance" events, might have led to a different history of penicillin.*

of the moldy culture itself than Fleming's habit of pursuing odd phenomena. He was accustomed to play and to studying "idle" curiosities. At first, he simply found the halo of inhibited bacterial growth *interesting*. Later in the day he toured the building trying to interest his colleagues—who were largely unimpressed. There was no promise of miracle cures yet. Later, Fleming drew pictures with *Penicillium* on culture plates and watched them "develop" over several days as the bacteria grew in the negative spaces. There was no "deductive reasoning." No "instant of great personal insight." Those elements seem to have been created for the sake of a heroic story.

Fleming in Historical Context

Nor did Fleming follow through on his own "discovery" in ways that someone who now knows the eventual importance of penicillin might expect. Fleming originally observed the action of penicillin in 1928. Yet he did not initiate clinical trials. Nor did he strongly advocate the use of penicillin in treating humans until 1940. The events during this twelve-year hiatus are perhaps the most telling in the history of penicillin.

Fleming was certainly searching for antibacterial agents in 1928. And he did investigate penicillin's potential. But he was not persuaded. He found that penicillin was not toxic to animals and that it did not harm white blood cells. Yet he also found that penicillin would not be absorbed if taken orally. Penicillin taken by injection, alternatively, was excreted in the urine in a matter of hours. For Fleming, penicillin's therapeutic potential was limited, perhaps to keeping superficial wounds free from infection.

Fleming did continue to use and advocate for penicillin in the years following his initial discovery. But the value he saw in penicillin was primarily in the context of bacteriology. Penicillin suppressed the growth of certain bacterial species, allowing him to selectively culture certain others (such as those causing influenza, acne, and whooping cough). In this role, penicillin became a valuable tool in the manufacture of vaccines—a major task Fleming managed at St. Mary's Hospital. Production of penicillin continued on a weekly basis throughout the 1930s, but all for purifying bacterial cultures. The penicillin was crude: good enough for Fleming's purpose, but hardly strong enough to destroy a serious human infection. Meanwhile, Fleming had turned his research to another group of chemical bactericides, the sulphonamides.

The Other Nobel Winners

The pursuit of penicillin in treating human infections was due ultimately to another lab, led by Howard Florey in Oxford. In 1938 Ernst Chain, an associate of Florey's, began a search for natural antibacterial agents, endeavoring to understand how they worked (Figure 21.3). He chose three to study, penicillin among them. Fleming's 1929 paper offered a thread of information that Chain could pick up, although with a quite different purpose in mind. By early 1939 Chain and Florey began to suspect the medical potential of penicillin. But they could not simply test it: penicillin was difficult to produce and to purify. Florey had difficulty finding funding. By that time, Britain was at war with Germany, and extra funds were not available for exploring mere possibilities. Support eventually came in late 1939 from the Rockefeller Foundation in the United States.

Florey shifted the resources of his department to the penicillin project. Before they could demonstrate the efficacy of penicillin, they had several technical challenges. They needed to improve extraction methods, refine an assay for determining the strength of their extracts, and scale up production. After five months of work, in May 1940 they had enough of the brown powder to test on mice. The penicillin allowed several mice injected with a lethal dose of virulent streptococci to survive. The potential of penicillin for treating infections then seemed demonstrably real. Florey and Chain repeated their tests as a double check and then went on to determine appropriate dosages and treatment durations, publishing their results that August.

FIGURE 21.3 *Others who helped in the long development of penicillin as a drug. Howard Florey (2nd from left) and Ernst Chain (4th from left) shared in the 1945 Nobel Prize in Medicine with Fleming.*

But the research was hardly done. Would the results transfer to humans? To determine that, they had to scale up production yet again. Based on relative weight, a human would need roughly three thousand times as much penicillin as a mouse. Commercial support was still not forthcoming. In the Oxford labs, flasks and biscuit tins used for the mold cultures gave way to hundreds of bedpan-like vessels stored on bookshelves. Purification turned from the laboratory to dairy equipment. Column chromatography allowed the group to isolate the relevant fractions and to concentrate their solutions. All this was in the service of an initial clinical test. After the first test, in early 1941, Florey's group had to return to their methods to find a way to remove some impurities that had caused side effects. The tests eventually went quite well, but producing enough penicillin to treat six patients had required two professors, five graduate students, and ten assistants working almost every day of the week for several months. Chain and Florey had recognized a therapeutic potential that Fleming saw only vaguely. And they had been willing to risk a substantial investment of resources to pursue it.

Fleming took notice of the striking results. But he did not disturb his research agenda. He knew that the value of penicillin still lay in inexpensive mass production.[6] Thus, the research—and, in a sense, the discovery—was still incomplete. Florey took his cause to America once again, where work began on the scale of breweries. One key technical assistant found a new medium for the mold cultures, increasing yields tenfold. Drug companies in England were by now interested, but the scale of production was at first somewhat limited. After a second set of clinical trials, in 1942–43, however, production began in earnest. In another half year, industry could produce enough penicillin to treat two hundred persons per month. Two years later, the United States was producing enough to treat a quarter million patients per month. If Fleming "changed the course of history,"[7] it was not without the help of Florey, Chain, and dozens, even hundreds, of technicians. Florey and Chain shared in the Nobel award with Fleming. The others did not.

Myth-Conceptions

Knowing the fuller history of penicillin, one may well reflect on the conventional tale of Fleming's "Eureka" moment. Why did Fleming's stature alone become so inflated? Here, focus shifts from history to the psychology of storytelling.[8] The typical account of Fleming and penicillin is not so much a glimpse into the practice of science as a tale meant to celebrate and inspire. Accordingly, fidelity to how science actually works becomes subsidiary. Accuracy yields to rendering great deeds by ennobled characters. So the hero's achievement is easily amplified and distorted.

First, elements of context are incorporated into actions and implicit intentions of the hero. Despite an emphasis on chance, Fleming ironically becomes the primary agent, with the stray mold in a supporting role. A casual observation is transformed into a more potent "great personal insight." All subsequent work is discounted. Others function merely to recognize Fleming's discovery "for what it was"—as if all the work were already done. The plot's action is propelled by the central hero and the pivotal "Eureka" moment.

Second, the collective contributions of many are collapsed into the triumph of one person. Fleming thus receives credit for having demonstrated the clinical efficacy of penicillin—actually the work of Florey, Chain, and others. The huge pharmaceutical industry seems merely to do his bidding. That's how important he was. The hero becomes monumentalized and thus exceptionally worthy—at the expense of others.

Third, the ordinary is transformed into the extraordinary, the mundane into the marvelous. The hero becomes larger than life. Fleming's original explorations over many weeks and months are thus reduced to one history-making moment. The boring work of trial and error is transformed into a stupendous "great personal insight." Heroic action is legendary and inspirational—precisely because it transcends human dimensions. Drama and emotion eclipse reality, with misleading consequences.

As a result of these tendencies, a complex set of circumstances, people, and events is molded into a tale with an all-too-familiar pattern. It is a good story—at least by rhetorical standards. It is engaging and appears informative. It has narrative tension and closure. It is inspiring. It is memorable. And so it is ripe for retelling, regardless of its truthfulness. Such is the psychology of storytelling.

Such an account is not a mere simplification. The story is noticeably (even if unconsciously) recrafted to meet certain aims. It is shaped to convey a larger-than-life image of how science happens through heroic deeds. The story thus purports to explain. Just as explanatory myths did in ancient cultures. One may thus call such an artificial construct a *myth-conception*. Such misleading stereotypes pervade popular stories of science (see also essays 22–24).

Ultimately, the conventional account of Fleming's "Eureka" moment is a morality tale. It functions to convey values through a narrative format. It tries to instill those values by drawing on the fame of the great discovery. It also endeavors to justify the values by situating them in an apparently inevitable and unchangeable past. A myth-conception thus constitutes an effort to *naturalize* an imagined ideal into the history of science (see essays 7 and 16–20).[9]

It is not easy to escape the psychological allure of such stories. However, one may begin by recognizing the narrative strategies at work. Awareness of the rhetoric helps expose the sacred bovine about "Eureka" moments of genius. Once liberated from that blinkered view, one can begin to understand authentic scientific practice. Beyond that, one might reflect not just on historical accuracy or the psychology of heroic myth-conceptions, but also on the very ethics of storytelling in science.

22

Round versus Wrinkled: Gregor Mendel as Icon

Round *versus* Wrinkled. It's like a coded message. Every biology student learns it. Pea seeds have one of these two shapes. The great botanist Gregor Mendel investigated how they are inherited, along with six other pairs of traits. In each case, traits do not blend. One is dominant, the other recessive. This is how genetics works. Mendel was smart. He chose the right organism. He chose the right traits. He counted. This is how science works. Foundational lessons in both genetics and the nature of science seem to emerge from this one simple case.

Yet the common lessons—a set of sacred bovines—are misinformed. Traits are not found mostly in Mendelian either-or pairs.[1] Dominance is not a representative genetic model.[2] Nor does scientific progress depend solely on genius and special insight. Mendel's method, like much research today, involved a bit of loose, open-ended searching. One can learn a lot about Mendel beyond the common, often-ritualistic retelling by focusing just on this one trait in peas. The widespread errors allow us to reflect on how we create our sacred bovines—and perpetuate them.

Either-or?

The traits themselves are nice examples of how genes work. Round and Wrinkled, we now know, result from the activity or inactivity of an enzyme in the pea seeds, starch-branching enzyme (SBE1, or SBEA).[3] SBE1 turns the pea's sugars into long-chained polymers of starch. That means fewer molecules. That in turn means the peas do not absorb much water (the process that biology textbooks duly label *osmosis*). Later, when the seeds dry, they also lose little water: hence, Round (or Smooth). When SBE1 is absent or not fully functional, sugars accumulate. More water is absorbed. Later, more water is lost and the seed coat shrivels: hence, Wrinkled. Because of the higher concentration of sugars, the wrinkled peas are also sweeter. By focusing on the sugars in the cell, rather than seed shape, we might alternatively label Mendel's Round versus Wrinkled as Starchy versus Sweet.

The Starchy/Sweet distinction is reflected further in the size and form of the starch grains in the pea cells. As R. P. Gregory observed in 1903, Round peas have

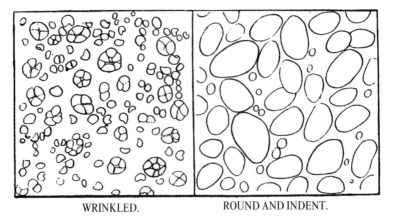

WRINKLED.   ROUND AND INDENT.

FIGURE 22.1 *Round versus Wrinkled. Or Starchy versus Sweet? Size and form of starch grains vary in pea seeds, helping to reveal that dominance is not absolute.*

large, simple starch grains, while Wrinkled peas have more numerous, smaller grains (Figure 22.1). The simple microscopic observation amplifies the question, what feature(s) properly defines the trait?

The size of the starch grains in peas was important to early geneticists. After the revival of Mendel's work in 1900, they avidly re-examined the seven traits that Mendel himself had studied. For example, A. D. Darbishire, of the Royal College of Science, noted that in hybrids of Round and Wrinkled (F_1 heterozygotes), the starch grains were *intermediate* in size and form. They also absorbed an *intermediate* amount of water.[4] In today's terms, an intermediate level of SBE1 synthesis yields an intermediate physiology. Mendel, of course, had identified Round as "dominant" and Wrinkled as "recessive," terms he introduced and that remain today. Yet here Round did not fully eclipse Wrinkled. Both traits seemed important, *even in hybrids.* One classic example of dominance, then, needed qualification, as noted by one of Mendel's greatest champions in Britain, William Bateson.[5] The case of the starch grains in peas, among others, led many early geneticists to question Mendel's implied principle of dominance.

Mendel's critics included, ironically, Bateson himself. Initially, he interpreted the appearance of intermediates—such as the starch grains—as exceptions to Mendelian inheritance. One may easily imagine, as many early geneticists did, that the hereditary information had "blended" irreversibly. Acknowledging his own error, Bateson clarified the concepts: "The degree of blending in the heterozygotes has nothing to do with the purity of the gametes."[6] In modern terms, expression of the genes is distinct from how they are inherited. Bateson did not thereby endorse any principle or law of dominance:

> In the *Pisum* cases the heterozygote normally exhibits only one of the allelomorphs [alternative phenotypic forms] clearly, which is therefore called the dominant. It is, however, clear from what we know of cross-breeding that

such exclusive exhibition of one allelomorph in its totality is by no means a universal phenomenon. Even in the pea it is not the case that the heterozygote always shows the dominant allelomorph as clearly and in the same intensity as the pure dominant.[7]

Even in the pea, Bateson noted. Mendel himself had noted many intermediate forms in *Pisum*. Just before introducing the concept of dominance, he observed, "with some of the more striking characters, those, for instance, which relate to the form and size of the leaves, the pubescence of the several parts, etc., the intermediate, indeed, is nearly always to be seen."[8] Mendel noted other exceptions: stem length (the hybrids were actually *longer*),[9] seed coat color (hybrids were more frequently spotted),[10] and peduncle length.[11] In *Phaseolus* (green beans), intermediate forms were seen in the hybrids of seed size, flower color, and seed coat color.[12] Bateson's later work on poultry echoed these cases. He showed that traits typically mixed in hybrids, even when they segregated and recombined in producing further offspring. Bateson concluded, "The supposition that dominance was an essential phenomenon of Mendelism was of course a delusion."[13] Bateson did not adopt dominance as a basic model, even while promoting Mendel's other ideas.

Other geneticists echoed Bateson's concerns. They denied dominance for many traits, even in Mendel's own peas. Mendel's Round/Wrinkled and other dominant/recessive trait pairs, they believed, did not fairly represent genetics. Erich Tschermak noted: "The appearance of the dominating or the recessive character is not an all or none phenomenon. In individual cases I was sometimes able to establish, with certainty, a simultaneous appearance of both, i.e. transition stages."[14] Carl Correns, one of the original rediscoverers of Mendel's work, also doubted any universal either-or principle: "I can not understand why DE VRIES assumes that in all pairs of characters which differentiate two strains, one member must always be dominant. Even in peas, where some characters completely conform to this rule, other character pairs are known, in which neither character is dominant, as for instance the color of the seed coat, being either reddish-orange or greenish hyaline."[15]

Dominance was not the norm. Thus, in 1926, when Thomas Hunt Morgan summarized the findings of the golden age of genetics in *The Theory of the Gene*, dominance had no primary role. Dominance is not critical or central to understanding heredity. Indeed, the concept itself is deeply problematic and fosters a wide range of misconceptions among both students and scientists.[16] Ironically, Mendel's peas—including Round/Wrinkled (or perhaps Starchy/Sweet)—are a prime example of the irrelevance of this "Mendelian" principle in heredity.

One may wonder, then, about the significance of dominance for Mendel's original study. If dominance is not typical, why did Mendel focus on such traits? Were they essential to reaching his conclusions? Popular accounts of Mendel convey only his renowned seven pairs of traits, implying controlled crosses between fourteen varieties. Yet Mendel worked with *twenty-two* true-breeding varieties. According to historians Vítězslav Orel and Federico Di Trocchio, he followed *at least fifteen* traits initially.[17] (Varieties would have differed by several traits, not just one.) Ultimately, as Mendel himself reported, he abandoned traits that were "difficult to define."[18]

That is, he explicitly excluded traits that "do not permit of a sharp and certain separation."[19] He relied instead on Round/Wrinkled and other such dichotomies. Accordingly, he qualified his law of hybrids to only "those differentiating characters, which admit of easy and certain recognition."[20] His decision to follow dominant/recessive pairs was not carefully planned in advance.

As Bateson noted later, the presence of intermediates does not interfere with observing segregating genes or even counting offspring precisely. Clear and distinct does not necessarily mean binary. Thus, Mendel could well have reached the same conclusions even without limiting his study to dominant/recessive pairs. Indeed, dominance obscures the relationship between appearance and genetic make-up. That is, *without* dominance, intermediate hybrids bear witness to genetic contributions from both parents. Mixed genetic make-up always looks different from pure genetic make-up. Genetic elements can thus be traced more transparently through the generations. When hybrids cross, the original traits still reappear in pure form in the subsequent generation. The blending of hereditary material is apparent only. That is how one infers that genes occur in pairs and that they separate and recombine each new generation. A 1:2:1 ratio is observed directly. It is not disguised as a 3:1 ratio. Paradoxically perhaps, Mendel's patterns seem easier to interpret using traits with no dominance (for example, starchiness or water absorption, rather than Round and Wrinkled).

The Limits of Simplicity

The apparent simplicity of the Round-versus-Wrinkled dichotomy can be deceptive in yet another way. Presenting Mendel's traits, echoed sevenfold, as a primary or fundamental model tends to imply that the typical trait has only one gene and that the typical gene has only two alternatives. Subsequent elaborations of inheritance in textbooks—significantly labeled "*non*-Mendelian"—only reinforce the image of what is "normal." Again, early geneticists underscored many exceptions to the simple Mendelian model even with Mendel's own traits. Seed shape, for example, seems influenced by at least one other gene. Bateson noted that indented peas "may be confounded with true wrinkled peas," even though they have large starch grains, like Round.[21] Currently, *five* loci are known to affect starch content, and each can produce wrinkled seeds.[22] Flower color and pod shape seem regulated by multiple genes, as well.[23] Mendel himself alluded to a third pod color beyond green and yellow, "a beautifully brownish-red."[24] He likely saw evidence of a separate pigment gene (*Pu*, for anthocyanin), rather than another allele of the same gene affecting chlorophyll. Mendel certainly presented the possibility of multiple controlling factors. Traits can have multiple genes, not just one.

The same multiplicity applies to character states. Round and Wrinkled reflect only two of many possibilities for SBE1. Bateson reported on "a great diversity both of rounds and of wrinkleds."[25] In molecular terms, subtle changes in the shape of the SBE1 protein (due to mutations and different amino acid sequences) will yield a range of enzymatic rates. Mendel focused on traits that he could easily differentiate visually. That facilitated counting—but at the cost of detailed realism.

Mendel himself learned the limits of his results on peas when he tried to extend them in his subsequent (now lesser known) studies on hawkweed, or *Hieracium*. There he could not detect the simple 1:2:1 pattern in offspring.[26] Students of Mendel's work need to be aware how choosing only certain traits limited his conclusions. Round versus Wrinkled can easily misrepresent genetics if not interpreted in the context of Mendel's selective method. The alternative is a misleading, even if unintended, impression of simplicity in nature (essay 21).

Popular accounts also tend to suggest that Mendel anticipated his findings and planned to test his hypothesis, as dictated by the so-called scientific method. However, his use of twenty-two varieties and his selective use of results indicate otherwise. According to Di Trocchio's reconstruction, Mendel began by searching broadly, documenting everything, not knowing what to expect. He probably crossed all the varieties with each other, hoping some significant result would manifest itself. By counting offspring, he would be able to express the number of each character type in each successive generation as a formula, or mathematical law of hybrids. He likely used partial results from crosses of plants differing in many traits, treating each trait independently of the others. For Di Trocchio, Mendel's traits are inherited independently precisely because he *disregarded* traits that were linked and whose numerical pattern he could not interpret. It is not mere coincidence that Mendel's seven traits are scattered across separate chromosomes, where they do not interact. Those traits likely emerged as an outcome of his experimental approach. Overall, then, Mendel seems to have followed a strategy of blind search and selection. Such a method is commonly adopted when scientists explore unknown complexity. Mendel's example, properly portrayed, may thus help us break the misleading stereotype of "the" scientific method and illustrate instead the "scientist's toolbox," with its many complementary methods (essay 5).

Romanticizing Stories

The deeper history offers several lessons about how science works and *how we tell stories about* how science works. For example, it is easy to understand the concept of preliminary exploratory investigation. Indeed, it tends to render scientific thinking as more familiar. The narrow, more formalized study that emerged later develops context, and its structure gains more meaning. Science is not so monolithic or rigid as the image of one method suggests. One can appreciate, too, how Mendel's results were selective. Mendel's model is informative through deliberate simplification. At the same time it is partly misleading, like most models in science. Mendel's insights emerged from ordinary reasoning. That means that in the right context and with appropriate background (and maybe a bit of luck), anyone might contribute to science. It is not the exclusive province of special people labeled "geniuses."

While we can recognize the misleading impressions in the conventional story of Mendel, it seems equally important to reflect on how such basic errors become so widespread and deeply entrenched. Mendel's work is typically rendered to fit a romantic image of Mendel as an idealized hero. The real Mendel is lost, replaced by a misleading caricature. An icon. Mendel is now the quintessential geneticist— and an ultimate authority. Historian Jan Sapp has shown how Mendel's status and prestige are so powerful that they can eclipse a fair or honest account of what he actually did or believed. He supposedly supported contradictory concepts, depending on whom you ask. Yet everyone appeals to Mendel's beliefs as the final arbiter. What Mendel did and believed are now the "gold standard." So you gain credibility if you can convince others that your ideas were once Mendel's ideas.[27]

Among educators, Mendel's work has been shoehorned to fit an imagined norm of "good science." His concept of dominance also earns undue importance. But would acknowledging Mendel's limitations really diminish the discovery or his greatness? The myth-conception of Mendel eclipses the authentic Mendel. It distorts understanding not only about the nature of science, but also about the very science itself. Sometimes, sacred bovines can trump plain evidence.

The most fascinating element of Round and Wrinkled, perhaps, is how widely it is known. Virtually anyone can describe it, even years after taking a biology class. Mendel is an almost universal rite of passage. Why is this historical episode special? Mendel's story has a powerful allure, not measured fully by the scale of his achievement. Contrary to the popular narratives, Mendel's Round versus Wrinkled is *not* simple. Dominance is *not* the norm. Not even Round and Wrinkled exhibit uniform dominance. Traits are not binary. Seed shape reflects *multiple* genes and *multiple* alleles. Mendel's method did *not* rely on special insight. He did *not* design a simple confirm/reject test. The simplicity of Mendel's study, as Bateson hinted a century ago, is an illusion. Ironically, it may very well be the ill-informed perception of simplicity that inspires geneticists and biology educators to revere Mendel—and to regard the misleading stories as sacred.

23

William Harvey and Capillaries

Circulation of the blood is so familiar that one can hardly imagine a time when it was not fully understood. Indeed, the ancients knew about the pulse and the flow of blood. They recognized, too, the vital importance of the heartbeat and nourishment. Yet the concept of a complete blood circuit emerged only in the early 1600s, largely owing to investigations by William Harvey (Figure 23.1). Harvey has since earned renown as one of biology's great heroes. But what guided Harvey to his landmark discovery?

According to many popular accounts, Harvey's genius was reflected in his remarkable ability to deduce circulation without being able to observe the capillaries that ultimately close the circuit between arteries and veins. Moreover, Harvey's reasoning was so powerful, they contend, that he was able to confidently *predict* the presence of the tiny blood vessels without ever seeing them. Only later did others confirm his insightful prediction. That triumph, tragically too late for Harvey himself to appreciate, seems to vividly demonstrate the importance of deduction and prediction in Harvey's work—and in science generally.

However, these stories do not measure up to historical evidence very well. Nonetheless, the widespread error is itself telling. Probing the erroneous stories more deeply, one can gain an even deeper appreciation of scientific myth-conceptions and how they foster misconceptions about the nature of science (essay 21).

Predicting Capillaries?

Most important to understanding Harvey's discovery, perhaps, is his adoption of the renewed spirit of experimentation in the early 1600s: an eagerness to tinker with and actively probe nature (essay 1). Rather than read books, he dissected animals. He cut open fish, frogs, and other creatures to observe their beating hearts. His unexpected observations led him to new conclusions, which he published in 1628 in *De Motu Cordis et Sanguinis*, or *On the Motion of the Heart and the Blood*. (Take quick note of the title: it indicates that the custom at that time was to use Latin in scholarly work—a subtle first clue to the cultural distance between then and now and the challenge of interpreting Harvey in a different historical context.)

FIGURE 23.1 *William Harvey, who discovered the circulation of the blood but not, as one might imagine, the capillaries that complete the circuit between arteries and veins.*

In the first half of his book, Harvey described the motion of the heart and arteries. They do not actively "breathe" and fill, like lungs, as the ancient Greek physician Galen and others had described. Rather, the heart contracts as it beats. At the same time, the arteries expanded from the influx of blood, in an *opposing* rhythm. Harvey described his observations of living animals (vivisection), especially of "lower" animals. With their slower hearts, the stages of motion were more readily seen. From observing blood flow in fishes and fetuses, and noting how heart valves work, Harvey reasoned all blood must flow from one side of the heart to the other via the lungs.

In the book's second half, Harvey argued for full circulation. First, an immense volume of blood passes through the heart from the veins to the arteries. Where does it all go, if not returning to the heart to cycle again? Second, a series of half-tight ligatures demonstrates that blood moves away from the heart and collects in the veins of the extremities. Third, valves direct the blood flow in the veins towards the heart only. Here, Harvey included a now-famous diagram showing how readers could demonstrate this for themselves (Figure 23.2). Evacuate the blood from a segment of vein in the forearm, then open the segment alternately at either end. Blood fills the vein again only from the end farthest from the heart. Taken together, these three observations, Harvey claimed, demonstrated "motion, as it were, in a circle."

The modern reader may notice that Harvey's argument seems to omit one key element. That is, blood must pass from the arteries to the veins. But how? Capillaries complete the circuit, yet without a microscope, Harvey could not possibly have observed them. Surely he must have understood their role even without seeing them. That is, he must have reached his conclusion using logic, deducing the need for capillaries from theory alone, and thus predicting something yet unknown. That

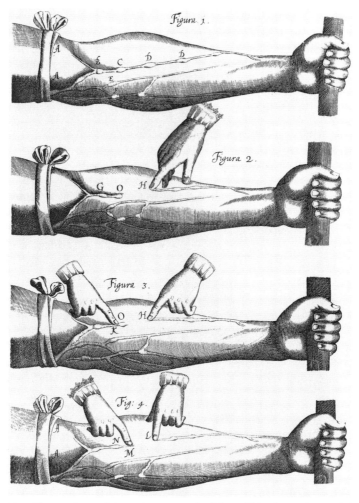

FIGURE 23.2 *Harvey observed that valves prevent backflow of blood, even in emptied veins, and thus that venous blood flows only toward the heart.*

leap of imagination, that bold conjecture, only seems to deepen Harvey's greatness. Both Marcello Malpighi and Anton von Leeuwenhoek observed capillaries several decades after Harvey's death, verifying the alleged prediction and bringing a satisfying sense of closure to the unfinished story.

Denying Capillaries

However plausible this view may seem, the historical documents indicate otherwise. Harvey did not theorize about capillaries in *De Motu Cordis* or any other publication.[1] To modern perspectives, this seems quite impossible. For us, how else could blood circulate?

One need only consult Harvey's writing. He said that blood "percolates" in the lungs. The veins absorb blood from the "pores and interstices" of the tissues. As an analogy he refers to "the way water percolating the earth produces springs and rivulets."[2] For Harvey, the tissues were like porous sponges that would yield their blood when in motion or when adjacent muscles contracted.[3] The rate of blood flow even varied through the organs, depending on their "denseness or sponginess."[4] For Harvey, the blood passed freely from arteries to veins. We observe this today in mollusks, arthropods, and other organisms. Biologists call them open circulatory systems. No blood vessels connect the arteries directly to the veins.

Ironically, perhaps, Harvey even argued *against* capillaries. Centuries earlier Galen had reasoned (deductively) that "anastomoses" (networked channels) must exist to allow blood to travel from the veins to the arteries. Harvey proposed the opposite direction. He was thus primed to discredit Galen's inference with observational data. Harvey reported many years later: "I myself have pursued this subject of the anastomosis with all the diligence I could command, and have given not a little both of time and labour to the inquiry; but I have never succeeded in tracing any connexion between the arteries and veins by a direct anastomosis of the orifices."[5] He boiled organs—the liver, lungs, spleen, and kidneys—until they were so brittle that their dust could be shaken from the fibers and he could trace every blood vessel distinctly. But he found no anastomoses.[6] Not only did Harvey not believe in capillaries; he looked for them and concluded they were not there.

Nor was the later discovery of capillaries even guided by consideration of Harvey's concept of circulation. Malpighi made his now-landmark observations while focusing on the structure of the lungs. In a letter to his mentor Alfonso Borelli, Malpighi noted the limits of unaided observation of the lungs:

> the blood, much divided, puts off its red color, and, carried round in a winding way, is poured out on all sides till at length it may reach the walls, the angles, and the absorbing branches of the veins.

> The power of the eye could not be extended further in the opened living animal, hence I had believed that this body of the blood breaks into the empty space, and is collected again by a gaping vessel and by the structure of the walls.

Echoing Harvey, Malpighi saw "empty space" between the observable blood vessels, where blood "poured out" and was "collected again by a gaping vessel." But using his microscope, Malpighi then observed "that the blood flows away through the tortuous vessels, that it is not poured into spaces but always works through tubules."[7] The observation of capillaries was quite unexpected, not due to any test of Harvey's ideas. Indeed, Malpighi did not refer to Harvey at all.[8] Stories that depict Malpighi as testing and vindicating Harvey's idea also fabricate history. Narratively, of course, the fictitious connection brings a satisfying sense of resolution. It thus helps contribute a purely emotional dimension to viewing prediction and testing as central to scientific discovery. The prejudices of storytelling trump a portrayal of how science truly works.

Learning from Historical Errors

At first, the error about Harvey and the "prediction" of capillaries may seem a minor historical footnote, of import only to a fussy historian. But the error—like any error, perhaps—holds great significance for the reflective observer. Namely, why did the error ever occur?

Consider, in particular, the vast reach of this error. One finds the story of Harvey predicting capillaries in a nationally syndicated radio series on science, a biographical reference from a university publisher, a prominent website, professional peer-reviewed articles for biology teachers, a public broadcasting series, and even Wikipedia.[9] One need not fault these sources to appreciate the impressive scope of the error. Even many historians once succumbed to the same mistake.[10] One cannot simply brush aside the error as trivial or due to willful ignorance. If errors are opportunities for learning, what does this particular error about Harvey indicate about how we think and tell stories—and thus how we might think more effectively in the future?

First, we are liable to read Harvey's seventeenth-century work in terms of today's science. For example, one may interpret Harvey's concept of circulation as our own. One may readily assume—without even recognizing that one is making an assumption—that Harvey accepted the capillary model. Any strange wording or odd-sounding phrases that might clue us to his different perspective—and they are frequent—may be easily dismissed as confused or irrelevant. Our minds subtly filter exceptions to match expectations.

One may also project one's own ideas about the nature of science onto history. In this case, the view that science necessarily follows a hypothetico-deductive format, often touted as the "scientific method," became the basis for interpreting Harvey's work. On this view, one far too readily imagines that Harvey was primarily a *theoretical* thinker and *deduced* capillaries. Historical facts are added and others trimmed to fit the presumed model of science. Again, the historical error might be inconsequential were it not for an added irony. Harvey's case has become for some a model example to parade before students showing that science is "essentially" hypothetico-deductive in nature.[11] What started as an assumption was written into, or naturalized in, history (essay 21). The history then seemed to justify that indirect inference as direct fact. Moreover, Harvey's fame contributes to making it an *important* fact. Through circular reasoning, the flawed historical interpretation leads to misunderstanding the nature of science.

A second theme also emerges when reflecting on the mistaken notion that Harvey discovered capillaries. That is, Harvey's achievement is inflated. This error, too, would matter far less were it not echoed further in depictions of Harvey's work. So, for example, Harvey was not the first to notice blood flow to and from the lungs. Miguel Serveto, Realdo Columbo, and Andrea Cesalpino each described the circuit of blood to the lungs in the decades before Harvey. (Arab physician al-Qurashi, known as Ibn al-Nafis, did likewise in the thirteenth century, although his work did not reach Renaissance Europe.) Likewise, the valves in the veins were first observed by Harvey's teacher Gerolamo Fabrizio (Fabricius). Harvey is portrayed as a progressive thinker in criticizing Galen, although others had also questioned Galen's centuries-old claims and authority. Like most scientists, Harvey worked in a tradition. Yet popular stories tend to attribute the separate discoveries to only one person.[12]

Once Harvey is cast as a singular hero, his own errors recede, his critics are vilified, and his triumph is amplified. For example, consider Harvey's analogies. In describing physiology, he alludes to cisterns, gunshots, and pistons. Some immediately conclude that he thereby established a modern, mechanical view of the body, including the heart as a pump. Yet Harvey ultimately attributed all motive power to the warmth of the blood. That view, now abandoned, remains hidden in most popular accounts. So, too, is Harvey's argument that the body is a microcosm of the world and the heart its sun. Harvey's mistaken belief in special life-giving forces also pervades his other great work, *On Generation*.

The drama is further heightened through opposing characters and conflict. Galen, in particular, is blamed for blatant errors perpetuated for centuries. Yet even Harvey himself explicitly praised Galen's expertise. Finally, Harvey's critics, such as Jean Riolan, are portrayed as inept fools impeding the progress of science. Their reasoning or evidence in historical context is dismissed. At its worst, science is reduced to a melodrama between heroes and villains. Scientists become superhuman legends—and at the same time, hollow cartoon characters.

The errors about Harvey, then, reflect the myth-conception syndrome also found in other popular histories (essays 21, 22, and 24). The actual history is not just edited for the sake of simplicity. The Harvey story actually *adds* the unsubstantiated facts about predicting capillaries and testing that prediction. These fables—masquerading as authentic history—convey only positive achievements, unfailing method, and idealized role models. An implicit moral about science emerges: proper method is logical and "by the book"; solutions are guaranteed; evidence is unequivocal; scientists transcend human limitations. Such stories may seem to inspire. But they also mislead. The story about capillaries and related versions of history are not idle. As myth-conceptions, they subvert effective understanding of the nature of science.

The Tragic Hero of Childbed Fever

The situation in the Vienna hospital in the mid-1840s was certainly grim. The hospital offered medical care to indigent mothers, but in one maternity ward women faced a ghastly one-in-ten chance of dying from childbed fever (today's puerperal sepsis, a bacterial infection). Could nothing be done?

Enter Ignaz Semmelweis (Figure 24.1), who, so popular stories typically tell us, "notices that [the attending medical] students move between the dissection room and the delivery room without washing their hands."[1] The students offering care are themselves infecting the patients with putrid matter from cadavers! Semmelweis institutes handwashing, and the mortality rate soon drops by an impressive 90%. However, "despite the dramatic reduction in the mortality rate in Semmelweis' ward, his colleagues and the greater medical community greeted his findings with hostility or dismissal."[2] Semmelweis's hypothesis "was largely ignored, rejected or ridiculed. He was dismissed from the hospital and harassed by the medical

FIGURE 24.1 *Ignaz Semmelweis, whose claims about the role of contaminated hands in spreading disease received a mixed reception.*

community in Vienna, which eventually forced him to move to Budapest."[3] Further injustice seemed to follow. "Despite various publications of results where hand-washing reduced mortality to below 1%, Semmelweis's observations conflicted with the established scientific and medical opinions of the time and his ideas were rejected by the medical community. Some doctors were offended at the suggestion that they should wash their hands and Semmelweis could offer no acceptable scien-tific explanation for his findings."[4] "The years of controversy and repeated rejection of his work by the medical community caused him to suffer a mental breakdown. Semmelweis died in 1865 in an Austrian mental institution. Some believe that his own death was ironically caused by puerperal sepsis," the very disease he had tried to prevent.[5] "Semmelweis saved the lives of countless women and their newborn children. He showed how a statistical approach to the problems of medicine could demolish popular but mystical theories of disease. His work prepared the way for Pasteur's elucidation of germ theory. He turned obstetrics into a respectable science. And he revealed how professional eminence and authority could breed crass stupid-ity and bitter jealousy."[6] Such is the apparently tragic tale of Ignaz Semmelweis, "savior of mothers" and "an unrewarded hero of medicine."[7]

The common story is a morality tale, where science offers prospective salva-tion, but fails. The sense of tragedy is poignant. And the emotion seems integral to the implicit lesson about the value of science, evidence, and open-mindedness versus prejudice and disdain. By evoking a sense of unjustified loss, the stories ren-der a set of unmet (as well as unstated) expectations. Others *should* have accepted Semmelweis's claims, because ultimately he was right. Others *should* have listened to him and honored him, because his scientific understanding could save lives. Heroic science, it seems, must have heroes, even if tragic ones. And villains, too. The heroes require ideal virtues, the villains the opposite. Yet these assumptions—these sacred bovines—can distort our understanding of the nature of science, by shaping how the very history is told. By delving a bit more fully into the history behind the stan-dard tropes of heroic tragedy, one can gain a deeper, less mythic, more authentic understanding of scientific practice.[8]

A Failure of Evidence?

First, if Semmelweis was right about handwashing, why was his work rejected? In conventional stories, the critics of Semmelweis were unreasonable. That is part of the genre. Anything less would diminish Semmelweis's heroic stature. In a sharp dichotomy, the evidence favors Semmelweis exclusively, while his critics were biased by "unscientific" factors. Their criticism apparently betrays personal interests and social ideology. Some stories say that Semmelweis was Hungarian and portray him as a victim of Austrian prejudice against foreigners. At the time, however, Vienna was viewed as "the Mecca of medicine." In an effort to reform the earlier excesses of bloodletting, purgatives, and other questionable treatments, doctors endorsed a doctrine of minimal therapy. "For fear of distorting symptoms, doctors refused to prescribe any remedies."[9] They were cautious in their conclusions. Without knowing a disease's cause, one could easily err. This was a valuable corrective. Doctors discarded artificial aids during childbirth, trusting nature to do its part. As such, the approach embodied the Hippocratic motto, "First do no harm." Our modern practices of diagnostics and loose bandaging of wounds began here. All this reflected a major advance in medicine. Semmelweis's claim about an ill-defined cause of childbed fever could justly prompt skepticism (even now, a hallmark of good science). Erroneous assumptions could divert one from finding the real cause.

Indeed, the stories often fail to mention that Semmelweis was *not* universally rejected. Many colleagues immediately adopted handwashing practice and advocated his views as far away as Paris and London. One journal editor compared the discovery to Jenner's smallpox inoculations in terms of importance. Semmelweis's results appeared in medical journals across Europe—in Austria, England, France, and Denmark. The practice of handwashing was adopted in England and much of Germany, at least. Semmelweis *did* help widely transform practice, lowering the incidence of childbed fever.

Yet Semmelweis also stridently expounded an explanation that went beyond the available evidence. He attributed *all* cases of the disease to decaying matter, even where none was evident or seemed plausible. While we may readily accept his view as championing our modern concept of a single common cause, observations at the time did not support this bold claim. Accordingly, most critics took exception to just this aspect of Semmelweis's ideas. One sympathetic reviewer noted that his claims "go too far and are too one-sided. In any case, Semmelweis owes us a proof that only the one etiological condition that he identifies is responsible. Nearly every obstetrician is still of the opinion that a large number of cases of illness remain that originate from a different cause, a cause admittedly yet unknown."[10] Doctors rejected Semmelweis's overstated theory even while adopting handwashing. In historical context, their selective judgment was based on evidence.

The need for washing hands seems obvious to us today because the concept of disease-causing germs is so familiar. That sense of ordinariness did not exist in the mid-nineteenth century. Physicians were certainly trying to isolate causes of diseases. They especially looked to anatomy for clues, correlating damaged structures with functional losses and symptoms. Yet the anatomical aspects of childbed fever were hardly uniform: why, with such diversity of effect, would one assume a single cause? Other theories looked at environmental factors and how diseases might spread via foul air, or miasmas. In Boston, renowned medical reformer Oliver Wendell Holmes, Sr. similarly tracked the contagion of childbed fever to physicians' handling of patients. But even he acknowledged a possible role for miasmas. In London, John Snow was facing similar challenges in persuading peers that cholera might be caused just by isolated sources of sewage-contaminated water. Semmelweis's explanation might fit, but in historical context the evidence was not complete or conclusive.

Calling the evidence unequivocal thus misportrays real, historically situated science.[11] But it does enhance the drama. Adversaries who deny "unequivocal" evidence generate emotional conflict. They amplify the hero's struggle. Such literary embellishments help transform ordinary history into a larger-than-life myth. A portrayal of authentic science is traded for emotional engagement. The case of childbed fever is indeed tragic, but perhaps only because truths do not announce themselves. Discovery is a historical process. And science's institutional norms for reliability foster caution. Handwashing practice could be adopted at the same time Semmelweis's theory was not because, ironically, evidence emerges piecemeal and science is inherently conservative.

A Role for Personality

Nor did Semmelweis always seem to help his own cause. One finds historical hints of a rather prickly and irascible personality. Semmelweis was not able to renew his temporary position at the Vienna hospital. One could blame the hospital director. He might have retaliated against Semmelweis for faulting programs he introduced, including adding dissections to the medical training and promoting the care of destitute patients for educational purposes. But documents indicate larger institutional power struggles. Furthermore, Semmelweis was obsessed with childbed fever. His passion may have eclipsed his other responsibilities. Neglect of some of his duties could easily have justified not extending his appointment. His zealous theoretical proclamations did not engender respect. Semmelweis ultimately decided to depart from Vienna but, mysteriously, left without even consulting or informing supportive friends.

Semmelweis also seemed reluctant to publish his findings, while expecting others to heed them. He acknowledged his dislike for writing, possibly rooted in unpleasant experiences in early schooling. His book appeared finally in 1861. Afterward he became increasingly intolerant of others who did not agree with him. He seems not to have tried resolving disagreements with critics or seeking aid from supporters. Rather, his unsolicited letters to critics were, as characterized modestly by one historian, "highly polemical and superlatively offensive." For example, to one German doctor he wrote, "I declare before God and the world that you are a murderer and the 'History of Childbed Fever' would not be unjust to you if it memorialized you as a medical Nero, in payment for having been the first to set himself against my life-saving theory." To another he alleged, "You, Herr Professor, have been a partner in this massacre.[12]

Ultimately, few popular stories mention Semmelweis's often-dismissive tone. That does not align with the heroic ideal. Instead, his critics, as implicit villains, receive all the blame. They are portrayed negatively, as biased by nationalism, hostility, pride, and pettiness. The asymmetry sharpens the sense of conflict—or of moralistic melodrama. It invites the reader to sympathize with the main character's struggle. But one cannot justly blame the critics for Semmelweis's own antagonism. Scientific consensus is inevitably social. Persuading peers is one of the challenges. Individual personality and demeanor matter. In retrospect, one can only note sadly the psychology that seemed to increasingly isolate Semmelweis as he aged, describing a tragic hero of another sort entirely.

Understanding Discovery

Ironically, many of the stories about Semmelweis do not profile very well how he made his noteworthy discovery—what may contribute most to understanding the nature of science. The problem of childbed fever was well known, even internationally. In some stories, however, one gets the impression that only Semmelweis-as-bold-hero tried to remedy it. The mortality rate in the Viennese maternity ward was also well known, especially when contrasted with the lower rate in the hospital's main ward. Nor was the difference between the wards—midwives in one and medical students in the other—a subtle secret awaiting notice. Yet this observation alone was hardly enough to indicate the source of the problem or its solution.

Semmelweis earns credit in part for his systematic investigation. He addressed multiple possible explanations in turn, subjecting each to rigorous scrutiny or testing. He considered birth delivery postures, the roughness of handling patients, overcrowding, ventilation, diet, the unlikely effect of clergy attending the dying, localized miasmas, and other possibilities. Semmelweis persisted, gradually discounting each as unsupported by evidence. He argued strongly from exceptions and statistics. His later test of handwashing, in fact, was meaningful in part because he could rule out these many alternatives. A story in which Semmelweis considers only his final hypothesis fosters an impression of immediate insight, or genius. That may endow the hero with exceptional intellect, but it misrepresents the process of science. Here, Semmelweis's sheer perseverance through mundane trial and error was significant.

Ultimately Semmelweis was clued through happenstance. As with the case of Fleming and penicillin (essay 21), his "discovery" might seem like a monumental insight if one disregards the many details that naturally threaded the thought process together. In Semmelweis's case, a colleague died unexpectedly. One can imagine the puzzle when an autopsy revealed pathologies resembling those Semmelweis already knew from the awful deaths of his maternity patients. He inferred that similar effects would be due to a similar cause, even though childbed fever was considered related to the uterus and thus to occur exclusively among women. Had his friend contracted the disease from a patient? With his regular but harmless exposure to such cases, that would seem unlikely. Semmelweis also learned that his friend had cut himself during an autopsy days before he died. He concurred with others that cadaverous material entering the wound had undoubtedly led to the illness. Taken together, these particular facts indicated that childbed fever, too, probably originated with the cadavers. That would explain the well-known difference between the two wards, since only the medical students (not the midwives in the second ward) participated in dissecting corpses. Following his methodical approach, Semmelweis then subjected the notion to further testing through handwashing. A fortunate, unplanned convergence of events showed Semmelweis—as it would for anyone—the key connection. There was no brute deduction. No rare insight of genius. Just individually fortuitous circumstances and commonplace reasoning.

Semmelweis did not know or speculate how putrid matter from corpses could cause disease. When he sought a way to rid the hands of the disease material, he relied on smell only. Washing hands in water, even with soap, was not enough. The odor remained. Only chlorinated lime solution seemed strong enough. While it proved effective, he could not say why. There was a limit to Semmelweis's understanding. That became clearer later, when germ theory emerged. There had been no hint of microorganisms behind the practice of handwashing. So implying, as some stories do, that Semmelweis anticipated or contributed to that very theory misportrays the historical context. Not even scientific heroes can transcend the limits of their time. But again, that impression would contribute to a sense that Semmelweis was "ahead of his time," and tragically unappreciated.

A Tragic Ending?

The final element in most stories about Semmelweis is the circumstances of his death. Semmelweis's outlook sadly deteriorated and developed into antisocial behavior that even his wife could not tolerate. Many stories try to relate this to the scientific tale by making the critics-as-villains directly responsible for the psychological decline. Whatever Semmelweis's own view—and his antipathies and anxieties have been noted above—this attribution is historically unwarranted. Eventually, his family and friends contrived to commit him to an asylum. But he resisted, was beaten, and died soon thereafter of a septic infection. Truly unfortunate. But in this case, the downfall and ironic closure, although not really relevant to the science, seem to many too tragic not to mention. Is it not sufficient to note merely that Semmelweis suffered a personal sense of neglect, never fully appreciating the value of his own impact?[13] Or does the rhetorical allure of the tragic hero demand more?

This essay has made clear throughout, I hope, that the story of the case of childbed fever has been shaped and widely retold, trimmed and falsely elaborated, all to cast Ignaz Semmelweis in the genre of a tragic hero. The history is sacrificed to achieve melodrama. But when the history is misleading, so too is the portrayal of the nature of science. Yet the real story is just as compelling. Evidence is not always unequivocal. Theories may be partly wrong at the same time they are partly right. Personality matters to persuasion, and to maintaining a scientific community. Discovery can arise from unplanned contingencies and ordinary reasoning, not just extraordinary insight.

From the Semmelweis case one may reflect more generally about historical stories in science. Heroes need not be perfect to be heroes. Indeed, how much more human and accessible they seem if we understand their flaws and limits, as well as their triumphs. The primary tragedy of this case may be failing to recognize that science, ironically, can often be effective without exhibiting what we imagine to be its essential ideals.

PART VII

Values and Biology Education

25

Respect for Life

"Respect for life" is a rallying cry among many animal rights activists. They target the use of animals in research, the wearing of fur coats, the hunting of wolves and whales, and more. They deplore, too, dissection in biology classes (essay 26).

Ironically, in many of these high-profile cases, "respect for life" seems to mean not a universal respect for all life, but rather a narrow "respect for life-like-us." The values seem associated primarily with mammals. A little knowledge of biology might broaden the scope of such sentiments, with perhaps some stunning consequences. We observe life in many forms. From cuddly kittens to slithering snakes and spooky spiders. From huge, awesome blue whales to microscopic, one-celled organisms. From lilies to "sea lilies," which are more closely related to starfish and sea urchins than to plants. And we are related to all of them. That is the startling lesson of evolution. We share ancestry with all these creatures. But how often is the grandeur in this view of life eclipsed by narrower feelings?[1]

For example, one hears much fuss over dissection in biology classes—of cats and frogs and fetal pigs. One hears very little objection to dissecting crickets, scallops, or worms. Why? While activists raid research laboratories where mammals are used, no wellspring of objection has emerged about a popular school lab exercise on the survival rates of sowbugs in different environmental conditions, although it assures wholesale death for some populations. Where is the hue and cry over bouquets of fresh-cut roses? Yet all are living.

Animal rights advocates frequently express outrage at killing animals for fur. But rarely do we see outrage over leather jackets, which are basically just fur hides without the hair. Leather shoes, leather belts, leather hats, leather gloves, leather bags, leather wallets, leather briefcases, leather watch straps, leather key fobs, leather whatever: there are animal hides everywhere. Shouldn't we be equally if not more outraged about the disparity of wealth between the persons who buy fur coats and the persons who, despite animal rights, want them but cannot afford them? Do we think about the lives of persons in war-torn and starvation-ridden regions with the same passion as the lives of furry animals?

How well do we reflect on respect for life? Consider, for example, the shelf of household insecticides (Figure 25.1): common weapons for killing roaches, ants, mosquitoes, wasps, ticks, and termites—all living things. How many people will

FIGURE 25.1 *Where is respect for life in household pesticides?*

campaign for animal rights while using such insecticides without a second thought? Where is our respect for the life of arthropods?

Consider also the pesticides for the garden, the cotton field, the wheat crop, or the orchard. How many people happily accept eating apples with blemishes on them? We don't want to think about harvest losses or, worse, developing alternative agricultural methods, and so we blind ourselves to billions upon billions of insect and plant deaths. Where is our principle of respect for life in pesticides?

Note, too, lawns and golf courses. Grass, a living thing, is grown for the express purpose of being trampled and cut back on a regular basis. For other species, that would be merciless exploitation. Not to mention the use of herbicides and widespread murder of dandelions and other broadleaf plants we choose to call "weeds." Where is respect for life in heavily managed lawns?

Plants remind us that living things encompass many kingdoms and that we might open our horizons still further. What are antibiotics for, but to eradicate whole populations of bacteria? Foot spray for athlete's foot, except to kill fungi? And disinfectant household cleaners, save for the genocide of our moneran and protist cousins? Where is our respect for microorganisms?

Finally, consider perhaps the most widespread use of animals in the United States: the raising in captivity of cats and dogs, not to mention hamsters, guinea pigs, fish, and parakeets. It is not enough that we deprive them of their liberty or breed them purely for our own enjoyment, apart from their once-native habitats. We as a nation feed them over 750,000 metric tons of meat and meat by-products every year.[2] Pet-food sales amount to over $23 billion annually.[3] Who else is not receiving food as a result? Does our "respect" for overfed pets not need reassessment when one-third of the world's human population goes hungry? Where is our respect for human life itself? Ultimately, what does "respect for life" mean?

Life covers the planet—even in the most unsuspected places. High atop mountains, where lichens cling to the rock surface. Deep in the dark ocean, where energy comes not from the sun but from the earth's mantle, through vents in the crust. In hot springs, where some bacteria accommodate boiling temperatures. Behind cave walls, where yet other bacteria can thrive by eating iron alone. Life is everywhere. Yet humans continue to expand their population, transforming habitats and reducing the space for other species to survive. Humans have filled the atmosphere with carbon dioxide and other greenhouse gases that have significantly changed climate patterns, threatening life forms globally.

Our modern culture owes itself some profound self-reflection on respect for life, it seems. Reflection about habitat reduction; sources of human food; medical research; recreational hunting and fishing; animal products; and the domination of pets in homes and animals in zoos; as well as insecticides, herbicides, disinfectant cleaners, and antibiotics. What might come of this reflection one can only guess. But it is hard to imagine that authentic respect for life will reduce to simple slogans.

26

Hands-Off Dissection?

Amid the mantra-like rhetoric of the value of "hands-on" learning, the growth of computer "alternatives" to dissection in biology education is a striking anomaly. Instead of touching and experiencing real organisms, students now encounter life as virtual images. Hands-on, perhaps, but on a keyboard instead. Or on a computer mouse, not the living kind.

This deep irony might prompt some to hastily redesign such alternatives. Or to find and adopt others. However, one could—far more deeply and profitably—view this as an occasion to reflect on the aims in teaching biology.

In What Sense Alternatives?

What do computer programs and models teach? By not sacrificing any animal, one ostensibly expresses respect for life. Nothing seems more important—or moral—for a biology student to learn. Yet using this standard—respect for life—many alternatives to dissection seem deeply flawed.

First, most alternatives share a fundamental destructive strategy of taking organisms apart. Each organ is removed and discarded in turn. That might seem to be the very nature of dissection. Yet some contend that "the best dissection is the one that makes the fewest cuts." Here, the aim is discovery, not destruction. One tries to separate and clarify anatomical structures: trace pathways, find boundaries, encounter connections—quite impossible if things are precut and disappear as preformed units in a single mouse click.

The "search and destroy" strategy, once common, is now justly condemned.[1] Such dissections were never well justified. They reflect poor educational goals and fundamentally foster disrespect toward animals. Indeed, dissections may be opportunities to monitor and thus guide student attitudes. Search-and-destroy *alternatives* to dissection merely echo antiquated approaches. Better *no dissections at all* than such ill-conceived alternatives.

Second, prepackaged images or take-apart models are not much better. They reduce the body to parts. No more than pieces in a mechanical clock.[2] They neatly parcel the body into discrete units. However, a real body is messy. It is held together with all sorts of connective tissue. Its compartments are lined by layers of membranes. There's fascia and fat. One body differs from another. That complexity just doesn't show up in textbooks or dissection programs that, in the name of education, simplify things. Demonstrating the gap between idealized textbook diagrams and reality is one extraordinary value of looking inside real organisms.

Here are some tasks that might guide a dissection oriented to discovering the body's organization:

- ▶ Trace back the path of urine, from a familiar point of excretion to its origin.
- ▶ Trace a piece of indigestible roughage from its ingestion to its excretion.
- ▶ Trace a molecule of glucose from its absorption in the gut to its use in a heart muscle cell.
- ▶ Trace a molecule of oxygen from the lung to the brain (or kidney).
- ▶ Trace the meeting of male and female gametes from each gonad.

No vague pointing! That's cheating. The tip of the probe must physically trace clearly visible structures. Now, that's "virtually" impossible with a computer program or a model.

A third inherent problem with alternatives to dissection is objectification. The animal isn't real. The virtual dissection is a game, just like other computer games. There is not even any basis for respect. Many critics of animal use focus not on the sacrifice of animals but on the psyche of the student. Invasive studies "can easily lead to insensitivity, callousness, and emotional hardening."[3] Such claims echo renowned philosopher William James and others, who alleged that animal experimentation habituated researchers to disregard pain and inured them to the suffering of humans as well.

Yet ghostly images on a screen and plastic replicas train students to respond to substitutes for living creatures. In *modeling* reality, they allow students to rehearse destructive actions. It is all the worse because there is no emotional engagement. It is, after all, *virtual* reality. What does it mean when they learn to butcher a body with no feeling? There is no sense of responsibility in interacting with something once living. When the virtual dissection is over, you close the program and erase any implications for the actions in that apparently unreal world. What you keep are the habits. Do alternatives educate students so that they can bomb villages in remote locations, secure behind a distant computer screen, with no human feeling? I fear that the emotional distance in computer programs and models may foster habits of a quite unintended kind.

Alternatives to dissection are ultimately often perverse alternatives. They tend to preserve the features of *inappropriate* dissections—destructiveness, reductionism, and objectification. Ironically, they do not teach respect for life.

Teaching Anatomy?

What *do* alternatives to dissection teach? Indeed, what *should* they teach? Many consider the goal of dissection as learning internal anatomy. Thus one study on dissection alternatives measured effectiveness solely in terms of anatomical identification.[4]

And why frog anatomy? Aren't we aiming to help students conceive of *them-selves* biologically? So if one is using virtual dissections, why use virtual frogs? Why not virtual humans? The frog (or rat or fetal pig) was generally meant as the best available substitute for the human anyway.

Encountering the insides of real humans, not just sterile diagrams in a text-book, might seem overwhelming. But it is fascinating and engaging, nonethe-less. In the European Renaissance, as social taboos about cadavers dissipated, human dissections spread publicly. Interest was hardly limited to anatomists. Vesalius held public demonstrations, as celebrated on the title page of his great treatise. Permanent anatomy theaters opened. Dissection became spectacle (Figure 26.1a).

Nor has fascination with the human body waned. Witness the enormous pop-ular response to Body Worlds, an exhibit of real human bodies preserved with polymer technology. Its creator, Gunther von Hagens, envisioned skateboarders, basketball players, dancers, and other bodies with their muscles fully exposed (Figure 26.1b). Of course, such exposure is dreadfully sensationalistic. Still, emo-tion can critically motivate learning. Perhaps that's what we need in the classroom instead of the skeleton in the closet?

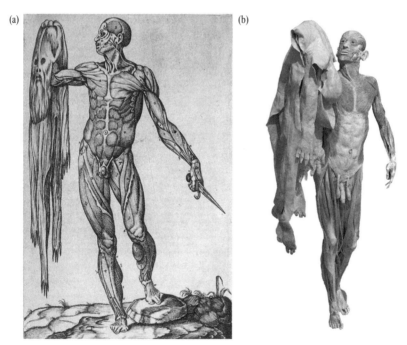

(a)   (b)

FIGURE 26.1 *Fascination with human anatomy, then and now.* Left: *A 1556 engraving by Juan Valverde de Amusco.* Right: *A modern reinterpretation by Gunther von Hagens, in the form of a plasticized body for public exhibition.*

Of course, one can begin to evoke the early anatomists' appreciation by sharing their magnificent drawings. Images from many of their great works are now available online (for example, see the National Library of Medicine exhibit at http://www.nlm.nih.gov/exhibition/historicalanatomies/home.html). In addition, through historical comparisons, one can visibly trace the scientific significance of investigative human dissection.

Renaissance authors on anatomy created their own version of virtual dissection. Reveling in the innovation of printing, they adopted a format now reserved mostly for children's books: foldouts! Images of the nude body included flaps. When lifted, they revealed the organs underneath—and then the organs underneath them. Alternatives to dissection apparently began quite early—for those *eager* for the knowledge.

If teaching simple anatomy is the goal, one may well question dissection as an appropriate strategy. One need not carve up an animal to learn where the organs are. One doesn't need an *imitation carving-up*, either. Old-fashioned diagrams work just fine. And sometimes, perhaps, just old diagrams.

Beyond Anatomy

But perhaps learning anatomy is not the aim? Consider how realistic images of the inside of a human body may disturb students. Why? They respond, surely, because *they* are humans. They are seeing themselves in an unfamiliar—and possibly uncomfortable—way. The anatomical knowledge here is irrelevant. Harm to animals is irrelevant. What matters is the relationship between the observer and the object being viewed. Yes, dissection can be deeply emotional: because we identify with the object being dissected.

Most students tend to view animals—and themselves—as black boxes. Encounters with the internal are rare. Surgery occurs isolated in operating rooms. Meat originates in remote slaughterhouses. Butchery occurs backstage at the grocery. Graphic images of war or terrorism or car accidents are (respectfully) omitted from broadcast news. Images that betray the tidy organismal black box can powerfully disrupt our psyches. Many people shudder to view bodies (implicitly their bodies) as assemblages of organs. Glimpses of violence on television and in film sometimes excite people—but largely because they violate an expected boundary. In such contexts, how can one learn the fundamental lesson about our very *biological* being?

Perhaps teachers need to also *personalize* the anatomy? For example, they could point to sites on their own bodies, and invite students to probe what's under their own skin. Do we fully value the tactile? Or does probing for one's spleen flirt with what may be another morally marginal territory: sensuality? (Note the potential irony that while scientists might underscore the role of sense data, in the classroom science teachers may fear entertaining the sense of touch—on one's own body, at least.) Yet even advanced students enjoy learning the skeleton through play: "Simon says, touch your clavicle. Simon says, touch your metatarsals. Touch your occipital lobe! Oh, Simon didn't say!" Here, at least, would be real hands-on learning.

But even such games or exercises fail to fully confront one's biological self. That level of understanding is deeply *emotional*. Opinions on dissection vary. Yet everyone acknowledges, I think, that what marks the experience for most is *affective*. The *feelings* tend to be primary. Quite so. Leonardo da Vinci wrote about such thoughts in his private notebook: "And if you should have a love for such things you might be prevented by loathing, and if that did not prevent you, you might be deterred by the fear of living in the night hours in the company of those corpses, quartered and flayed and horrible to see."[5] Dissection is not about teaching anatomy. It is about learning to address these profound emotions and, ultimately, to understand oneself.

Several emotion-laden elements of one's biological self are encountered in dissection. Opening the body's black box is the first. Students often invest considerable attention to the exterior of their bodies (their grooming, their muscles, their soft tissue), but when do they consider their physiological interior? Dissection offers surprises. Yes, the texture of the muscles is like meat: that's because that's what meat is. The texture of the brain is like a firm custard. (There are educational benefits from using freshly dead specimens.) The emotions are directly linked to self-understanding.

Second is awareness of the body's "messiness," as noted above. An idealized textbook diagram may be a tool for learning anatomy, perhaps, but not for appreciating the complexity and particularity of real bodies. This, too, has a deeply emotional component.

Finally, the whole experience is filled with the shadow of mortality. A full understanding of life is incomplete without a complementary understanding of death and dying. A challenge is addressing the emotional overtones of that topic gently and respectfully, but also with honest realism. William Harvey, in studying the heart and blood flow, performed many vivisection experiments (essay 23). That may seem callous and impersonal. Yet Harvey also attended the autopsy of his own father. And when his wife's pet parrot, regarded with much affection, died, he dissected it, too. Imagine the emotional context: even when dealing with death personally, Harvey expressed a profound respect for life. The emotional overtones of dissection are not necessarily ideally suppressed or disregarded. Rather, they may be integral to deepening an appreciation for life—including, notably, one's own life.

The history of anatomy offers another potential lesson here. In medieval Europe, human anatomy was taught in part using annual dissections. But they were largely demonstrations, following the ancient texts of Galen. Anatomical discoveries were few. The Renaissance "hands-on" attitude changed that. Dissection shifted more to active investigation. New structural and functional knowledge emerged.[6] By comparison, then, modern students might benefit from immersive experience—and not in order to learn anatomy. So, one may add to the list of dissection activities:

▶   Explore. Find something not on the list and discuss why you found it meaningful.

Many current alternatives to dissection, it seems, fail to address the most important lessons. The very awkwardness of dissection may be its touchstone. Models, whether plastic or pixelated, are not real. Indeed, they distance the viewer from the real body. The emotional interaction—which virtually defines the whole learning experience—is absent. In that sense, they are not alternatives at all.

Oddly enough, then, dissections—far from being an exemplar of disrespect—may actually contribute significantly to students' developing a deeper appreciation for life. This was certainly true in a college class I taught for nonmajors. In a poll one term, 95% of the students indicated that they learned something valuable from our rat dissection. That included feeling a deeper respect for living things. "Things are so much different than the way they are pictured and described in the textbook!" one student noted. "Diagrams give you just one side of the picture and sometimes even two. But with the dissection you get to see the organs in all dimensions, you get to look at it any which way you want and what's better is that you get to touch it. What I learned from the dissection is to appreciate what every single organ does for the body (me)."

In a culture where anyone who ate meat also helped butcher it and anyone who wore leather also helped skin the animal, dissection in a classroom would not be needed. Those experiences might be considered the real alternatives. Dissection may thus also be about learning the scope of one's actions regarding animals and thus understanding the corresponding responsibilities (essay 25).

Dissection may be paradoxical in its role in teaching respect for life by coupling informative images and emotional encounter. A successful dissection is one that someone would never need experience again, but without which a vital lesson might never be learned.

27

Organisms, Modified, Genetically

GMOs. Genetically modified organisms. They conjure the specter of "Frankenfoods." Monstrous creations reflecting human hubris. Violations of nature. And their very unnaturalness alone seems reason to reject the whole technology.

But one may challenge this sacred bovine: the common image that GMOs cross some new threshold, dramatically changing how humans relate to nature. Or even that such a view can properly inform how we assess the value or risks of GMOs. Rather, biologically, GMOs are modest variants. As I will elaborate, "conventional" corn is probably more deeply shaped by human intervention than any addition of, say, a single *Bt* gene for a pesticide-resistant protein. Many crops promoted as "natural" alternatives are themselves dramatically modified genetically, like the cats and dogs we enjoy as pets. And *this* perspective—the context of GMOs—should inform views on policy. Without resolving the question of ultimate risks, we should at least recognize and dismiss as irrelevant the claim that GMOs are "unnatural."

Modifying Organisms

While criticisms of GMOs vary, one recurrent theme is the assertion—or the implicit assumption—that they are inherently unnatural. For example, one high school student commented in a class discussion on genetically modified salmon, "Even though it definitely has many economic benefits, I think that shaping the way in which other organisms grow and live is not something that we as humans should be taking into our own hands."[1] As rendered recently for young readers, a cartoon princess of the Guardian Princess Alliance scolds a grower of GMOs: "These fruits and vegetables are not natural."[2] Many seem to believe that for humans to alter something living is to thereby taint it. Organisms should remain "pure." Nature seems to exhibit its own self-justified purpose, not to be disrupted.

What does this mean for all the other ways that humans modify organisms from their "natural" state? For example, we adorn our skin with tattoos and pierce various body parts. In certain cultures, at certain times, we have bound feet and elongated skulls. We may reconstruct our bodies "cosmetically" to suit our tastes, perhaps dissatisfied with what "nature" has provided. Indeed, most consider it a gift to create a smile where there was once a cleft lip.

We also fill teeth with metal amalgams and high-tech ceramics. We use orthotics to reshape bone structures. We transfuse blood. We replace body parts—transplanting kidneys, hearts, livers, corneas, and bone marrow. In some cases, we install human-made body parts instead: hips, knees, pacemakers, stents—more inorganic plastic, metal, and ceramics.

Nor do we limit ourselves to macroscopic physical changes. We inject synthetic hormones and immune suppressors. We ingest antibiotics. We deploy a large repertoire of chemicals that alter neural function: stimulants and depressants, pain relievers, behavior-modifying medications, and psychotropic drugs just for recreation. Perhaps these ways of modifying organisms are so familiar now that we fail to recognize them as human interventions. Or we may rationalize that because humans themselves evolved, these modifications are expressions of nature, too: of nature beyond nature.

Yes, genetically engineered organisms are modified. But they are modest "DNA transplants"—results of molecular surgery. Given the scale of entire genomes, at tens of thousands of genes, how significant are they? While some modifications yield measurable economic consequences, organically they seem quite narrow in scope. In other contexts, would we even notice the addition of one or two genes among so many?

Modified Genetically

For many, of course, the concern is not modifying nature per se, but specifically modifying *genes*. Genetic technology seems to threaten the integrity of a species. The very label "transgenic" may conjure images of fearsome hybrids, like the monsters of mythology: harpies, sphinxes, satyrs, manticores, and chimeras. One student asked rhetorically, "Are the [GM] crops still the same as before?" But genes are not identity (essay 19). In this case, a fear of genetic change reflects beliefs in biological essentialism—a vague conviction that species have a purpose just as they are now, apart from the perpetual flux of evolution.

Again, are we blinded to the familiar? We are surrounded by genetically modified organisms. For example, household dogs (Figure 27.1). Shepherds to shih-tzus, dachshunds to Dalmatians to Dobermans, bulldogs to bull terriers, they are all engineered variants of wolf-like ancestors, which were domesticated around fifteen thousand years ago. Since then, humans have generated specific genetic types through selective breeding. While dogs differ from wolves genetically by only about 1%, their diversity illustrates how effectively humans have already modified organisms—genetically.

The same applies, of course, to domestic cats. And domesticated cattle. Chickens. Pigs. Rabbits. Sheep. Goats. The variants of humanly engineered pigeons, from carrier to tumbler to pouter to fantail to capuchin, were certainly well known to Darwin, who commented on them in describing the powerful role of human selection in the first chapter of the *Origin of Species*. Humans have been modifying animals genetically for millennia, including many species that now contribute to our diet.

FIGURE 27.1 *Domesticated dogs all reflect human genetic modification.*

Domesticated species notably include food crops, as well. Kale, broccoli, cabbage, cauliflower, Brussels sprouts, and kohlrabi are all variants created by human intervention from one species. Bread wheat is a hexaploid hybrid of *three* wild species that humans developed about ten thousand years ago (Figure 27.2). Corn (maize) was developed from wild teosinte around the same time. Many of its historical genetic changes have now been identified, including genes for a softer kernel covering and for less branching in stem growth. There have also been genetic modifications to the "types and amounts of starch production; ability to grow in different climates and types of soil; length and number of kernel rows; kernel size, shape, and color; and resistance to pests."[3] "Conventional" corn is itself greatly modified genetically.

Why, then, is so much attention given to crops modified by a single gene—for example, the *Bt* gene? Humans merely moved the gene from a nonengineered bacterium, which had been applied externally as a pesticide, to the cells inside corn plants, where production and exposure of the *Bt* protein could be more circumscribed. In a historical context, the genetic modifications opened by molecular techniques are narrowly focused and relatively modest. Imagine how historical awareness might benefit a student who noted (not atypically), "I personally feel like the genes of organisms, especially those we consume, should not be tampered with."[4] Views of GMOs could well be informed by a deeper understanding of the role of humans in the genetic history of "conventional" crops.

The popular impression is also that scientists have the "unnatural" power to leverage any genetic modification and create any monster imaginable. Few persons reflect on how any organism needs physiological integrity and how this severely limits arbitrary changes. Not all hybrids are viable. Functional parts cannot be mixed and matched at will, except in fantasy. Hybridization and horizontal genetic

T. urartu	Ae. speltoides	Ae. tauschii	T. monococcum
$(A^u A^u)$	(SS)	(DD)	$(AmAm)$

T. turgidum ssp. dicoccoides	T. turgidum ssp. durum	T. aestivum
$(AABB)$	$(AABB)$	$(AABBDD)$

FIGURE 27.2 *Modern bread wheat (*Triticum aestivum, *lower right) is genetically modified— combining chromosome sets from three species.*

transfer certainly occur far more widely in nature than scientists once thought. (Ironically, nature itself seems to have violated what we once considered "natural" patterns!) Still, not every combination is developmentally stable. Nature still seems to limit what genetic changes are possible.

When new technologies are introduced, initial caution generally seems warranted. Yet genetic modification of organisms began decades ago, with microorganisms. Indeed, engineering bacteria to produce insulin, human growth hormone, interferon, various pharmaceuticals, and other products now seems almost commonplace. The basic gene-splicing methods are hardly new. Fuller awareness of this history of biology might also inform views on whether we are now altering nature in some way that is more "unnatural" than before.

Another expressed concern about GMOs is the possibility of "new" allergens. This potential risk, too, seems based on the image of GMOs as "unnatural," or inherently outside the norm. Yet allergens exist in all kinds of foods, not just GMOs—nuts, milk, wheat, eggs, and so on. GMOs are not new in this regard, either. (Indeed, governmental safeguards against such allergens in foods already exist.) The newness is overstated.

Ultimately, impressions that GMOs are genetically "unnatural" seem to foster all kinds of misleading perceptions, including about what risks may be relevant or significant. We seem to owe ourselves some reflection on genetic essentialism and "naturalness."

The Science of "Unnaturalness"

Science, it might seem, can solve the problem by describing objectively what is "natural" and what is not. Indeed, through observations and experiment, science can document what *is* or *can be* the case. Yet "natural" has a second, quite different meaning when discussing GMOs: namely, what *was intended* or *ought to be* the case. This requires a different kind of justification, based on values or assumptions about purpose. So, to contend that something is "unnatural," or aberrant, and thus *wrong morally*, goes well beyond what a scientist could validate. Ethical or teleological arguments differ from the epistemic arguments of science. The two types of "natural" claims are thus not interchangeable. Science is limited in what it can say about the "naturalness" of GMOs.

Still, science can help us reflect on the psychology of appeals to naturalness. Images of purposeful "nature" seem easy to come by. Humans tend to project their perspectives onto the world, conflating their personal interpretations with the way nature actually functions. They tend to assume that others will see it their way, too. The concept of "unnatural" thus seems based on an unspecifiable (but apparently "obvious") personal intuition. The vagueness certainly makes public discourse problematic. Views of "naturalness" also tend to become quite potent cognitively.

Emotions and attitudes can strongly shape an individual's "science-based" views of GMOs, as with other topics.[5] Our minds readily cherry-pick scientific perspectives to accord with prior beliefs or emotions. Thus, passionate advocates may accept flawed research uncritically. They may continue to cite outdated, discredited studies. They may discount counterevidence. Yet they may well still perceive their positions as thoroughly scientific.

The view of GMOs as "unnatural" or disvalued also contributes to a set of scientific misconceptions. For example, many people regard "natural" foods as inherently *better*, and *thus also* more nutritious. They presume that GMOs are less healthy, even when genetic modifications do not involve nutrients. They advance such claims even without scientific study. GMOs also seem less viable or less able to reproduce, because they seem less "authentic." At the same time, GMOs can apparently grow out of control and overtake other plants, even whole ecosystems, like nonnative or invasive species. The role of ideology is also indicated in the assumption that GMOs are able to "infect" other systems with their unnaturalness. (In the same way, many people infer that irradiated foods exhibit radioactivity themselves.) One student blogged, "If the genes in the salmon are genetically modified, who knows what it will do to your genes?"[6] Such a concern does not seem based on scientific understanding of how it might pose a plausible risk. No evidence indicates that other genes can produce such ill effects—only vague negative impressions.

Of course, anyone with access to the Internet could expose all these mistakes through their own GMO mythbusting. But the deeper question, from a cognitive perspective, is, why do such myths emerge at all? Here, these misconceptions can all be traced to a common assumption: that GMOs are "unnatural." Informal reasoning trumps science and evidence. As a result, arguments based on scientific evidence rarely penetrate these perspectives.

To address these types of errors, therefore, one must shift to another, more fundamental cognitive level. One must understand how perceptions of GMO "unnaturalness" originate psychologically. Generally, what we deem natural is "what we find more familiar, while what we consider unnatural tends to be more novel—perceptually and experientially unfamiliar—and complex, meaning that more cognitive effort is required to understand it."[7] In practical terms, then, "unnatural" means that GMOs are strange and hard to understand. The situation is ripe for feelings of insecurity and fear. Views about inherent purpose and genetic essentialism, discussed above, also seem to contribute. For scientific evidence about GMOs to have standing or persuasive merit, then, one may first need to engage the foundational views about "unnaturalness" and what they mean.

Consider the recent challenge of one county council member in Hawaii, faced with a proposed ban on GMOs there. The community overwhelmingly supported a ban. Local scientists and farmers did not. Taking his charge seriously, he tried to investigate the various claims. He had to learn some basic biology. He had to assess the credibility and expertise of his sources. It took time. But thoroughness helped him sort the quick rhetoric from more complete and reliable reasoning. He gradually distinguished the emotional hyperbole from the mundane but more systematic evidence. His eventual decision to vote against the ban was well informed, although not politically popular.[8] His case offers an occasion for personal reflection. Why do we believe what we believe about GMOs? Is it well informed by science? How do emotions and informal thinking shape our impressions of that information? And how do we effectively identify and manage the different sources of our thoughts?

Are GMOs "unnatural"? Who says so? In what context? Based on what background examples? One might even ask whether farming itself, as practiced through intensive monoculture, is itself "natural." The overall beneficial status of agriculture can itself be questioned.[9] Such an analysis, along with the other broad biological perspectives presented here, might help contextualize and inform the often-contentious issue of GMOs. Can we accept the risk that through critical reflection we might develop a fuller and possibly transformative understanding of organisms, modified, genetically? OMG.

28

Close to Nature

"Appreciation of life" is not listed on many biology curricula nowadays. Yet no other theme seems more central, nor so intensely personal, for many research biologists, as well as for many biology teachers (essay 1). Why exclude something so meaningful?

Of course, biology is a science. And science is about facts, not values. Besides, how would one evaluate a student's level of "appreciation" objectively on a multiple-choice test? Still, disregarding this strong motivational factor seems puzzling in an era when many teachers regard their primary challenge as trying to motivate students. Meanwhile, major institutions are trying to motivate more students and recruit them into science careers.

Perhaps we can address this puzzle? First, let's publicly acknowledge this dimension of research and teaching and begin to articulate why it is so important. Second, let's discuss how to teach this responsibly and perhaps how to assess the effectiveness of our teaching.

The National Association of Biology Teachers, for example, currently has position statements on the spread of AIDS, global warming, population control, and the use of animals in education. All reflect indirectly the general principle "respect for life." Yet it is not stated explicitly. Why not? Can it be assumed? Our culture embraces professional ethics. Should a statement of the values of professional biology educators include "striving to convey an appreciation of and respect for life"?

A passion for life seems most conspicuous when teachers address the topics of tropical rainforests, biodiversity, and environmentalism. Teachers often exhort their students to be concerned citizens, even to take public action. Some will promote vegetarianism. Others will encourage students to campaign for preserving species. And so on. While these practices embody principles of academic freedom, the introduction of particular ideology is also troublesome. It too readily mixes science and values in a way that can easily confuse students, who often struggle with the fundamental difference between descriptive and normative claims. Scientists can be experts, but they alone cannot tell us what *values* to follow, what we *ought* to do. Scientific arguments alone cannot secure biodiversity as a value, for example, although I fear many present them so. Surely, scientific knowledge informs us, but it is the basic values of long-term self-preservation and respect for life that support the biodiversity banner *morally*. Science teachers and scientists alike should be aware of the limits of scientific authority in the realm of values. Political advocacy is not the solution.

So, outright instruction on "appropriate" values is not legitimate. By what approach, then, might one justifiably teach appreciation of or respect for life? Students cannot appreciate something deeply without also understanding it fully. Developing a basic conceptual awareness of life is critical and foremost. Simply knowing about the ecological role of plants and microorganisms in our own survival (essay 10), for example, engenders a form of valuing them. Plants transform our carbon dioxide waste back into oxygen. Microorganisms convert our organic waste back into the basic nutrients that can be absorbed again by plants. The core value is nothing more than our own life. But other organisms are integral to sustaining that life. Understanding that relationship matters. Of course, ecological interconnectedness and systems thinking *do* appear as concepts on national and state biology content standards. So there is an opening, as long as we frame the knowledge in terms of its contextual meaning.

One might also consider the role of pure aesthetics. Consider the following experience. You enter your customary biology classroom, but today it is dark. An unseen source casts some gentle blue light on the ceiling, undulating across the surface. A few images of whales appear on the front screen, including a dramatic series of a humpback breaching. Meanwhile, the darkened room is permeated with eerie, ethereal sounds—deep stutters and wailing glissandos. Not a word. The whales' otherworldly songs fill the space, rolling and groaning, until, 45 minutes later, a class bell interrupts the soothing serenity. Students respond differently to such a class period. Some are puzzled. Some are annoyed. Others find deep appreciation. They have developed a sense of another organism, of feeling close to nature. That is a meaningful lesson, perhaps achieved in no other way. The experience of those sounds was certainly important to Roger Payne, a researcher in bioacoustics. His fascination with the whale sounds led him to decipher their meaning: as songs, sung in repeated cycles by males during mating season.

Experiencing the intricacies of life processes, experiencing the startling diversity of life itself, or experiencing the passion of research can all deepen an appreciation of life and of the scientific enterprise.[1] In addition, experiencing the unexpected can open our eyes, as well as our feelings. Life often exhibits more potential and meaning than we at first imagine. Keeping an eye out for the new and unusual has its rewards.

For myself, appreciation of life is found in viewing lichens (Figure 28.1). Lichens are favorite examples of symbioses—those close associations of species that can illustrate the web of life. A fungus and an alga, with two wholly different ways of securing nutrition, live together with mutual benefits. The symbiosis occurs unseen, at a molecular level. Lichens help remind us that humans are atypically large as organisms go. To understand life in general, including our own lives, we must often change our perspective profoundly.

FIGURE 28.1 *Lichen in Sandbjerg, Denmark—biology as an aesthetic experience.*

Lichens live minimalistically, at the margins. On otherwise lifeless rock, they find what is essential to survive. In otherwise barren environments, they take hold. Even where it is cold and dry, they sustain themselves. Yet lichens retreat when air quality diminishes, faithful indicators of environmental quality. At once resilient and fragile.

Lichens reflect an age-old theme of complements: chance and necessity. By necessity, they grow where they can harvest light and essential mineral nutrients. By chance, they grow in patches on a rocky substrate, scattered across the landscape. One can observe this pattern visually, but only at the appropriate scale. When the field of view is too large, lichens become insignificant blotches. Too small, and the parts overwhelm the whole. Appreciation is based on a particular way of seeing and knowing. That opening into a new understanding is what learning can provide. And it may reflect a fundamental, if often unarticulated, goal in a meaningful education.

Once upon a time, education aimed to foster an "appreciation" of the subject. That had two meanings at once. It signified, first, an intellectual understanding. But it also included an aesthetic feeling. Nowadays, the standard view of textbook-as-curriculum leaves little room for emotions. But perhaps we can find a place for just experiencing and enjoying a broad spectrum of life's processes. Appreciating phenomena as small as a colorful patch of lichens (Figure 28.1), or as grand as planet that "breathes" with the seasons (essay 2). Might we rediscover a place for wonder in biology education?

Indeed, if one looks at lichens closely and patiently, with an informed eye, sensitive to their patterns, one can become attuned to their reality. One may well sense oneself dissolving into nature. Immersed in wonder, one may learn to appreciate living things more deeply.

Epilogue: Challenging Sacred Bovines, Fostering Creativity

Where has this journey through misplaced assumptions in biology led us? From the heritage of Karl Marx's politics in studying cell organelles (essay 3) to the history of modifying organisms genetically before molecular engineering (essay 27); from heroic stories about conquering childbed fever in nineteenth-century Vienna (essay 24) to the errors of the great Charles Darwin and many Nobel Prize winners (essays 11and 12); from the wonders of monsters and human anatomy in the late Renaissance (essays 1 and 26) to a fascination with lichens today (essay 28); from hermaphrodites and twins and conjoined twins (essays 16, 17, and 19) to peppered moths, atmospheric carbon dioxide, and garden peas (essays 2, 20, and 22).

At one level, fuller awareness can inform our choices as consumers and our decisions as citizens. Savvy to the wiles of science con artists and more knowledgeable about the social architecture of trust (essays 13–15), we might be better equipped to assess the credibility of information about socioscientific issues. Aware of the complexities of scientific methods and funding (essays 2, 4, 5, 10, and 21), we can interpret more fully the practice behind scientific claims. Understanding the politics of the august banner of science and of heroic tales of scientists (essays 21–24), we can be more critical of inflated rhetoric designed to garner our allegiance. Likewise, we can more discerning about the pretensions of naturalized values—about genetic essentialism, or competition, or simplicity, or gender, or normality, or so-called human nature (essays 7–9, 16–20, and 27). Attuned to the consequences of our values and our prejudices, we may be able to express a more expansive respect for life (essay 25). We can be more informed where science intersects our lives.

As an ensemble, the essays here also indicate on a deeper level the pervasiveness of misplaced assumptions about biology in our culture. They seem to hide everywhere, sometimes in plain sight. They haunt our discourse and thinking about fingerprints as emblematic of identity (essay 19), images of competition as "natural" or inevitable on the sports field or in Congress (essay 7), the normality of a "male/female" checkbox on a health form (essay 16), the very concept of biological normality (essay 17), expectations of genius in science (essays 10–12, 21–24), claims denying vaccine safety or climate change (essay 15), or the extreme either-or nature of debates on the fluoridation of water supplies or HIV testing (essays 15 and 20). Much escapes our day-to-day notice. One might thus be impressed with the prevalence of our many blind spots.

Surrounded by sacred bovines that tend to remain invisible, one may well appreciate the virtues of being more mindful. Can we learn to be more perceptive? Can we foster habits for taking notice of and questioning assumptions? Can we become more-discerning individuals? Can we develop strategies for finding and remedying our own errors?

In the prologue, I introduced the evolutionary irony that our cognitive functions do not always serve us as we imagine. We have many blind spots. They make us susceptible to mistaken or misleading conclusions. The many essays here hopefully have made this deficit abundantly clear. At the same time, however, I noted how our minds are flexible. We can learn. We can develop new patterns of thinking to monitor and guide our unconscious processes and thereby catch our mistakes. The details from the many cases of sacred bovines also give us hope of, and perhaps clues to, learning how to challenge assumptions.

Learning to Challenge Assumptions

Nowadays many folks celebrate "thinking outside the box." It is a metaphor for challenging assumptions. Break free of convention. Get beyond the familiar and the mundane. Discover something new and exciting. Open new worlds. Be innovative.

The metaphor of a box is a fascinating image. It suggests a physical limit to our thoughts: that somehow, thinkers can be confined "inside" a box. As if the skull imprisoned our minds. The implicit goal is to free our thoughts. Abandon all assumptions. Escape limits. Eliminate constraints. Liberate our minds. Yet the nature of the box itself remains unclear. So creativity remains mysterious: the enigmatic product of undefinable genius.

The image of thinking outside the box may be misleading, however. All thinking has a context. There is always a perspective. There are always assumptions. There is always a box. We cannot think outside one. Ultimately, we can only think inside a *different* box. Or inside several other boxes. We can compare them and choose one. The essence of challenging assumptions is not getting rid of the oppressiveness of boxes. It is finding another box, appropriate for the occasion.

A first step, then, may be to understand the box one is thinking in. Or even to understand that one is in a box. As the many cases in this book indicate, this self-realization does not happen on its own. Assumptions tend to get buried. The box itself and its distinctive features are usually hidden. The first challenge—perhaps the most formidable—is to recognize any assumption *as* an assumption.

One educational aim already widely touted and pursued is *critical thinking*. It may seem a powerful antidote: one should simply discredit and reject ill-founded assumptions. Alas, "critical thinking" too often reduces to criticism—unfocused fault-finding. It inspires an oppositional stance to just about anything, merely for the sake of doing so. Accordingly, this approach typically yields a negative perspective, rather than a viable alternative. It engenders conflict, rather than insight. What one needs, by contrast, is openness to new alternatives. Instead of delving for deficits, one should seek novel perspectives. A fruitful posture is to be receptive, not antagonistic. Ironically, perhaps, one may need to *withhold judgment* to appreciate a different point of view. One needs to listen patiently and sympathetically, rather than search for flaws or expound easy criticism.

As an initial step, then, an open, reflective posture can enable one to acknowledge assumptions as assumptions. That is, any scientific claim embodies a particular perspective. Another perspective may be conceivable. One must be ready to entertain that other possibility, at least as an exploratory exercise. If the new alternative is well founded, we may then wish to explicitly change our working assumptions. The many essays in this volume, I trust, illustrate through multiple examples the value of engaging such possibilities, however strange or unlikely they may seem at the outset.

Yet our minds are habitually resilient. Novel possibilities rarely announce themselves. One must be on the lookout for them. We have to deliberately cross-check our familiar ways of thinking. We have to think twice, in a sense. And we may need mindful practice for this to become habit.

Another strategy commonly advocated to limit error, especially in science, is *skepticism*. That is, one should, by default, doubt any claim. Acceptance comes only after examining the evidence. Unfortunately, a generalized skeptical attitude may be no more productive than criticism. As noted in essay 14, trust is integral to our modern culture of specialized knowledge. Blind, broad skepticism eliminates too much. To be productive, concern about questionable evidence must focus on particulars. Reasons for doubt should be specific. Ideally, one does not dismiss an assumption without simultaneously having a clear reason for assuming something else. That is, one needs an active *analytical* perspective, not the hollowness of mere skepticism.

The clues to alternative ways of thinking emerge at the margins. Through exceptions that nudge the edge of our awareness. In expectations that may be disrupted. Anomalies—features that do not fit what we anticipate—may either frustrate us or puzzle us. Rather than discount them, we can profit by probing them a bit further. What do the discrepancies mean? Do they indicate an undetected assumption? Everyday ironies—the subtle inconsistencies we encounter as we go about life—are perhaps our most valuable tool for revealing misplaced assumptions. That is why they form the core of this book. They signal an occasion to reflect—and to appreciate the way we make assumptions.

The complementary task is to envision alternatives. It involves *taking perspectives*. As a skill, it can be nurtured. Again, the essays here offer numerous examples. The many cases in this book can be models for inspiration. From anomalies, one proceeds to "what if" scenarios. It is a form of play to wonder, "How might things be otherwise?" Reserve prejudicial judgment. In a provisional spirit, adopt a different interpretation. Entertain possibilities. Explore consequences. Most people can imagine alternatives—once encouraged to do so. Too often, however, the invitation is lacking. How might we benefit from more leisure time and a sense of security, to let our minds wander?[1]

A helpful cognitive ability is *making connections*. Drawing from our personal experience, we notice and follow similarities. Using those associations as cues, we resituate the case at hand in a wholly different context. This is how we find a new box. The new box is a different perspective, a fresh set of circumstances. The new juxtaposition structures suggestive "what if" conditions. That is, the initial connections guide analogies and metaphoric thinking. With the simultaneous similarities and differences, we can consider new relationships, new possibilities. Analogies and metaphors have long been recognized as important to creative thinking.[2] They help guide us "outside the box." Ultimately, new contexts help expose former assumptions and their meanings. Our sacred bovines become a little less sacred, perhaps.

Challenging assumptions is not solely an intellectual activity, however. Emotions enter as well. Familiarity offers a sense of security. Known quantities, even if wrong, are predictable and manageable. New assumptions lead to strange new ways of doing things. And also to fearful unknowns. So the innovator must be ready to *tolerate uncertainty* or periods of ambiguity. Novel assumptions lead to change, and the mere process of changing is not always easy. Confidence and emotional buoyancy help carry us through the unsettling feeling of chaos during mental reorganization. The cases in this book, I hope, exemplify how even the unimaginable might become reasonable.

In addition, changing core assumptions has a social dimension, with its own emotions. Disturbing the status quo can threaten the social order. Human cultures generally promote conformity to the group. Accordingly, once a thought pattern becomes established, the social context helps entrench it. Disagreement is discouraged. Dissenters may thus be counseled, "Don't rock the boat" or "Don't upset the apple cart." The social context contributes significantly to sacred bovines' becoming "sacred." Challenging them thus can disturb one's social well-being. So a further dimension of challenging assumptions is nurturing the *courage to disagree*.[3]

A knack for challenging assumptions, with its implications for creative thinking, is not the special province of rare genius.[4] It emerges from a set of skills that anyone can learn and develop. The relevant practices include (as profiled above) withholding judgment, adopting an analytical posture, reflecting on anomalies, respecting unfamiliar perspectives, making connections, playing with new ideas, exploring consequences, tolerating uncertainty, and risking dissent. We can all learn to challenge sacred bovines and, at the same time, to become more creative thinkers.

Beyond Assumptions: The Art of Learning and Creativity

The process of effectively challenging misplaced assumptions is ultimately a form of learning. We reach a deeper understanding of the world. That connection between analyzing cultural assumptions and learning may flirt with yet another sacred bovine: the view that learning is a cumulative process, based on simply acquiring more independent bits of knowledge. Most psychologists now agree that learning often involves displacing one concept with another. We may rebuild conceptual networks. Ideas become connected in new ways. New patterns emerge. Gestalts switch.[5] In the classroom, teachers thus ideally help students identify their preconceptions, introduce them to discrepant events that challenge the limits of those concepts, and then guide them through additional information and reflection to transform those initial concepts into something more sophisticated, more congruent with a larger reality. Learning often entails conceptual change or replacing one concept with another. In a sense, every act of genuine learning—of discovering something new—involves challenging an earlier assumption.

The lessons about how to challenge assumptions are thus not restricted to various misleading perspectives that haunt our culture. Nor are the lessons of "thinking outside the box" limited to what we might conventionally regard as creative work, whether in the arts or the sciences. They are the very skills for *learning how to learn*. An exploration of sacred bovines is ultimately a form of nurturing our capacities to learn anything.

Perhaps this should not surprise us. Independent learning involves the ability to diagnose and remedy one's own errors. The error-correcting process is fundamentally what challenging assumptions is all about. Our minds may be plagued with blind spots, as profiled in the prologue. But the journey through the sacred bovines in this book essentially maps the solution. It exemplifies the style of innovative thinking whereby we accommodate or mend those blind spots.

Questioning sacred bovines, I hope this book has demonstrated, can be engaging and enlightening. Fascinating lessons await just beyond the boundaries of commonplace assumptions. The world of biology is far more interesting than many biology courses convey. The process of science is far more complex—and compelling—than depicted in science textbooks or popular media. But even more than that, an exploration of sacred bovines can open a path to the satisfying reward of lifelong learning.

Afterword for Educators: Sacred Bovines in the Classroom

Most essays in this volume were adapted from a column for *American Biology Teacher*. This reflects the fact that while the themes are of general interest and importance, they are especially significant in the biology classroom and other educational settings—museums, nature centers, television documentaries, and other media formats. There, misplaced assumptions acquire an additional aura of authority, amplifying their ability to mislead. Once a sacred bovine gains a foothold in a textbook or in the culture of teachers, it is, unfortunately, easily echoed by others. Educators, ironically, often help perpetuate the entrenched errors. But they are also ideally situated to help remedy them. There is much *unteaching* to do.

First, I hope all biology teachers can become aware of the sacred bovines profiled in this book and appreciate, in particular, how they can adversely affect a general understanding of science and shape personal and social decision-making. Errors of naturalizing cultural values (essays 16–20) can have potent political consequences. Knowing what or whom to trust about scientific claims (13–15) and how science intersects with values (25–28) matters to well-informed public policies and personal consumer choices. Images of the scientific process (1–5, 10–12) and expectations of scientists (21–24) can ultimately influence the general credibility of scientific claims and the support and funding of the scientific enterprise. Teachers knowledgeable about sacred bovines can be more mindful in their lessons and in noticing and correcting common misperceptions. They can contribute more effectively to scientific literacy.

The essays here are not only for teachers, however. They are also designed to engage students. They are short. The cases are typically surprising and dramatic. They are informative as well as entertaining. They are a potential teaching resource in the classroom, whether for reading or presenting. Each essay is a small lesson in the nature of science.[1]

But the essays are more still. Another widespread goal in education is to teach critical and creative thinking. As noted in the epilogue, the essays here also function on this deeper level by demonstrating how to challenge assumptions and how to explore alternative perspectives. They are models for improving analytical thinking skills. Teaching such perspective-taking and thinking skills is widely acknowledged as valuable, yet methods for teaching them tend to be elusive. Too often, critical thinking reduces to just criticism. We want to foster more-productive thinking patterns. Students need examples to follow. The essays show not only how to notice misplaced assumptions, but also how to resolve problems by delving deeper and discovering new ways of seeing. Exploring sacred bovines can thus be a general educational tool for improving how our minds function.

Such creative approaches are essential, of course, to science itself. Every scientific discovery reflects, in a sense, an earlier blind spot, a bit of something we did not know previously. Scientists are human and thus susceptible to all the ways the mind fails to work optimally. With continued scrutiny, however, erroneous concepts and theories are encountered and eventually replaced. New findings can contradict old conclusions. We justly celebrate such scientific discoveries. Their novelty can sometimes astound us. We no longer believe that human "monsters" are supernatural omens (essays 1 and 17) or that tool use or morality are unique to humans (9, 18). We no longer accept the concepts of bodily humors, or of vitalistic forces in cellular chemistry or in embryonic development, or of stress as the primary cause of ulcers. Scientists are champions of challenging misplaced assumptions and developing new ideas. Teaching about sacred bovines, then, is also good training for becoming a scientist and for applying scientific reasoning in everyday life.

Science education reforms in the past two decades have underscored the importance of teaching more than science content. They have stressed a role for understanding the nature of science, "ideas about science," or "scientific practices"—or, more plainly, how science works.[2] By profiling biological errors and their correction, these essays can be valuable tools in that educational effort.

In a similar way, we often want students to learn how to make their own discoveries and revise their own knowledge. In one view of education, we teach what is known. In another view, we teach each person how to gain knowledge on one's own. How will students cope when they encounter what is now unknown or whatever we cannot teach them? Ideally, they learn *how* to learn. Again, the essays here can help prepare students by exhibiting some of the basic thinking processes involved (see the epilogue).

Even more than that, the numerous examples of sacred bovines—of misplaced assumptions—indicate the very importance of such learning. Some of these errors hide in plain sight, relatively immune to our perception. We can be seriously mistaken, even when we do not know it. Accordingly, we might adopt a more modest posture about what we think we know. Too often, it seems, we assume that we already know everything that might be worth knowing. Not so. In particular, by reflecting on the cases in this book, one can foster an appreciation of the complexity lurking behind apparently simple cases (see especially essays 4, 16, 20, 21, and 25). Simple stories or explanations can be grossly misleading. That is a lesson about the value of learning itself.

In summary, this collection of essays can contribute to teaching at multiple levels: (1) correcting misconceptions about biology and science; (2) fostering skills in error analysis and creative thinking; (3) developing thinking skills for science and independent learning; and (4) appreciating more fully the role and value of education itself. Challenging sacred bovines should ideally be a part of every biology curriculum.

Strategies for Teaching

Teaching about sacred bovines poses special challenges for teachers. Much of our educational system is founded on tactics for "delivering" preformed knowledge. State curricula often appear as lists of concepts to master. Standardized tests ask students to identify fragments of factual knowledge. The school culture and styles of classroom teaching have largely adapted to this environment. However, the lessons here are quite different in nature. What matters instead are ways of perceiving and thinking. What particular skills are required of a teacher venturing into this territory of teaching about sacred bovines and creative-thinking skills?

Learning is a form of conceptual change. And conceptual change, cognitive scientists tell us, involves rewiring the brain. New neural connections and pathways are established. We do not just stuff facts into empty storage holes. So, effective learning focuses on the transformation, not just the desired final outcome. That is, teachers need to guide students *from* preconception *to* reconception. Educators refer to this as a constructivist pedagogical strategy, and it is now widely acknowledged as foundational to effective, long-term learning. Students need to squarely confront anomalies or inherent contradictions and notice explicitly how they do not fit with older views. Again, there is much *unteaching* to do.

Then students need to explore the new ideas and envision how they "work" differently from the former familiar ones. Adopting this constructivist pattern, each essay in this volume begins by tapping into a commonly held belief. Only then are the ironies or counterexamples revealed to challenge the naive views. Subsequent comments introduce new views and examples, helping to replace the original misplaced assumptions with deeper knowledge. Each essay is a guided reflection, ripe for use in the classroom.

A teacher might present the cases here in at least two possible ways. The more familiar route may be for students to read an essay and then perhaps reflect on it individually by writing an entry in a journal. The reflection is essential. Students need time to consider the new ideas, especially to compare them with old ideas and accommodate any mental reorganization. Discussion with others adds more. Ironically perhaps, self-reflection is enhanced by interaction. The teacher, of course, as a facilitator and guide, is positioned to help resolve those discussions. That is, the effectiveness of reading by oneself is limited. Lessons need to actively engage students and then be formally consolidated.

An alternative strategy for teaching about sacred bovines is for the teacher to present the case, using images to help students visualize the strange cases or new examples. Here, the essay functions more as a "script" or implicit lesson plan for the teacher. Again, on constructivist pedagogical principles, one cannot just lecture about sacred bovines and expect to see any effect. Teachers need to involve students in actively considering the ideas. So a chief advantage of an in-class presentation is that the teacher can pause at any point, pose provocative questions to students, and engage them in discussion. Teachers can help contextualize the details and dramatize the ironies for their particular students. In this way, the lessons unfold through a series of cognitive steps, each with its own explicit reflection. Teachers can more closely and personally guide the process of conceptual change. Classes become more lively, through student interaction and active thinking, and provide much more enjoyment—and reward—for the teacher, as well as for the student. For additional support, teachers will find a sampling of visual resources and some prepared presentations online at http://sacredbovines.net.

More-ambitious teachers may be eager to capitalize on the opportunity, noted above, to also nurture more-creative thinking among their students. One approach is to intercede in class discussion whenever a particular component skill becomes relevant and to take note of it explicitly. As articulated in the Epilogue, these features include

- postponing judgment;
- adopting an analytical posture;
- reflecting on anomalies;
- respecting unfamiliar perspectives;
- making connections;
- playing with new ideas;
- exploring consequences;
- tolerating uncertainty; and
- risking dissent.

A teacher's observation can give the moment more weight. Again, the ultimate objective is not so much to enable students to label these skills as to foster new behavioral dispositions and habits of mind. Students should be invited to envision the alternative patterns of thinking and reflect on how that can open new ways of seeing things for themselves. One may regard this as a sort of stop-action rehearsal for applying new cognitive skills. Practice makes perfect. In this way, recurring discussion of sacred bovines can contribute to lessons beyond the conceptual content of each individual case.

These skills can then be reinforced on occasions when students are involved in their own classroom inquiries—whether of contemporary cases, historical cases, or student-based investigations.[3] Historical cases, in particular, can provide valuable retrospective insight, as exemplified throughout this volume. In inquiry mode, students can be challenged by the same kinds of problems addressed by famous scientists from the past.[4] This allows them to apply their creativity while developing skills in scientific practices, as recommended in at least one recent major document on science education reform.[5]

Yet another layer of learning can occur when students have been exposed to several sacred bovines (perhaps on separate occasions) and can think about them as an ensemble. Reflection can open awareness of the potential for one's own errors and analysis of how they occur. In the Prologue I noted several problematic challenges in human cognition: availability bias, confirmation bias, and entrenchment. Our minds seem prone to errors, as much as we also have the ability to find and fix those errors. Repeated exploration of sacred bovines helps open awareness at a general level of the importance of alternative perspectives. Accordingly, we might, first, gradually train ourselves to be less susceptible to the availability bias and to always be on the lookout for ways of thinking other than the first one that presents itself.

Second, we can be cognizant of the contexts that tend to reinforce faulty notions. We can learn the value of reassessing what may at first seem obvious or patently beyond question. That is, we can begin to confront confirmation bias by nurturing habits of entertaining different views. Finally, we can battle entrenchment with awareness of how even misplaced assumptions can become "sacred." Ultimately, we can learn to reflect more actively and think more creatively. All these elements can become explicit under a teacher's guidance. Again, with due investment in reflection and discussion, especially after exploring multiple cases, sacred bovines can be occasions for learning about cognitive errors and how to think more deeply.

Envoi

To echo the opening of the Prologue, what seems more obvious to the ordinary person than male and female as natural categories? Or that because genes govern our heritable traits, all the way down to cellular processes, they thus define our identity? Or that Darwin's principle of "survival of the fittest" aptly describes our fundamental selfishness in a competitive society? Or that biology typically advances by leaps of creative genius, epitomized by the rare achievements of our scientific heroes? Yet they are all misplaced assumptions. Biology classrooms offer an opportunity to expose them as ill informed. The world is far more complex and fascinating than is generally admitted. Even more, however, under the guidance of a good teacher who poses well-framed questions, an exploration of such sacred bovines can lead students to more-insightful and more-fruitful ways of thinking. And that would be a triumph in science education well worth celebrating.

ACKNOWLEDGMENTS

Most of the essays here appeared first as part of the "Sacred Bovines" column for *American Biology Teacher*. I am deeply indebted to the editors, Ann MacKenzie, Bill Leonard, and Bill McComas for their support over the years.

I have benefitted immensely from the close and critical eye of colleagues who volunteered to review the essays, especially Robert Dennison and Alex Werth. I am also thankful to Robert Cooper, Ami Friedman, Sherrie Lyons, Eric Howe, Alice Dreger, Joan Roughgarden, Phyllis Novikoff, Elaine Challacombe, Tim Dennis, Arthur Knowles, and Elizabeth Henderson.

Publication support—of a conspicuously professional quality—has come from Cheryl Merrill, Richard Earles, and Mark Penrose. I have equally enjoyed collaboration with and guidance from fellow members of the National Association of Biology Teachers: Dan Wivagg (co-author on essay 5), Maura Flannery, Rita Hoots, and Jose Vasquez, among others.

More deeply, I want to acknowledge key individuals who have supported my role in adopting contrarian perspectives, and who have helped me develop a distinctive "voice," or writing style: Dan Oates; Bruce Ruble; Michael Kirchberg; my parents, Nancy and Richard Allchin; Marjorie Rachlin; and Joy B. Osborne.

In addition, I gratefully acknowledge permission to republish the essays from their original sources. Essay 4 is adapted from "Appreciating Classic Experiments," in Carolyn Schofield (Ed.), *2004–2005 Professional Development for AP Biology*, College Board (New York, 2004). Essay 20 is adapted from "Kettlewell's Missing Evidence, a Study in Black and White," *Journal of College Science Teaching, 31*, 240–245 (2001). Essay 24 is adapted in part from a letter to the *Journal of College Science Teaching* (October 1, 2000). Versions of essays 21 and 25 appeared in the *SHiPS Teachers Network News*. The analyses presented in essays 20, 21, 22, and 24 echo those presented earlier in "Scientific Myth-Conceptions," *Science Education, 87*, 329–351 (2003). Essay 25 is based on the presentation "Teaching Respect for Life" at the Symposium on Dissection, sponsored by the Humane Society of the United States, October 16, 1996, Charlotte, NC. I am grateful to Jonathan Balcombe for the invitation to participate on that occasion. The remaining essays appeared originally in *American Biology Teacher* from 2002 to 2014. My special thanks to Jacki Pepin of the National Association of Biology Teachers for her gracious encouragement in republishing them independently.

NOTES

Prologue: The Ironies of Misplaced Assumptions

1. Kahneman (2011).
2. Gilovich (1991); Nickerson (1998); Sutherland (1992).
3. Wimsatt (2007).
4. Campbell (1974); Plotkin (1994); Wimsatt (2007, pp. 75–93).
5. Cooper and Carlsmith (2001); Harmon-Jones and Mills (1999); Piaget (1971).

1. Monsters and Marvels

1. Hertel (2001).
2. Hertel (2001, p. 9).
3. Daston and Park (2001, p. 14).
4. Daston and Park (2001); Della Porta (1658); Smith and Findlen (2002).
5. Purcell and Gould (1986, 1992); Rumpf and Beekman (1999); Musch and Willmann (2001).
6. McCusick (2004).
7. Hertel (2001, p. 17).
8. Smith and Findlen (2002, p. 1).
9. Musch and Willmann (2001).
10. Paré (1573/1982).
11. Daston and Park (2001, p. 154).
12. Daston and Park (2001, p. 285).
13. Findlen (2002, pp. 306–310).
14. Paré (1573/1982, p. 38).
15. Daston and Park (2001, p. 316).
16. Blumberg (2009); Bondison (1997, 2004); Leroi (2003); Purcell (1997).
17. McManus (2002).

2. Ahead of the Curve

1. Leaf (2011).
2. Quinion (2011).

3. Marxism and Cell Biology

1. Allchin (2007b); D. R. Holmes (1989).
2. Novikoff (1945a, p. 215).
3. Noviokff (1945a, pp. 213–214).

4. Novikoff and Holtzman (1970).

5. Novikoff (1945a, p. 214).

6. Novikoff (1945a, p. 211).

7. Novikoff (1945b).

8. Novikoff (1945b, p. 93).

4. The Messy Story behind the Most Beautiful Experiment in Biology

1. DNA Learning Center (2011); F. L. Holmes (2001).

2. F. L. Holmes (2001); Meselson and Stahl (1958).

3. F. L. Holmes (2001, p. 368).

4. F. L. Holmes (2001, p. 429).

5. F. L. Holmes (2001).

6. Delbrück (1954).

5. The Dogma of "the" Scientific Method

1. Watson (1968).

2. Di Trocchio (1991).

3. Allen (1978).

4. Mayr (1991).

5. Grant and Grant (2002); Weiner (1994).

6. Bazerman (1988); Knorr-Cetina (1984); Medawar (1964).

7. Bauer (1992).

8. Anand (2002).

9. Einstein (1954, p. 283).

6. Was Darwin a Social Darwinist?

1. Huxley (1894/1989); Richards (1987); Ruse (1986).

2. Stent (1978).

3. Bartal, Decety, and Mason (2011); Ganguli (2006); Langford et al. (2006); L. J. Martin et al. (2015).

4. Bradie (1994); Farber (1994).

5. Wright (1994, pp. 287, 310).

6. Browne (1996, 2003).

7. Darwin (1871, p. 169).

8. Barrett et al. (1987).

9. Darwin (1859, p. 488).

10. Darwin (1871, p. 70).

11. Rottschaefer (1998); Sober and Wilson (1998).

12. For recent relevant work, see de Waal (2009); de Waal and Suchak (2010); Hauser (2006).

13. Darwin (1871, pp. 71–72).

14. Darwin (1871, pp. 73, 92, 97, 98). For modern perspectives, see Clutton-Brock and Parker (1995).

15. Darwin (1871, pp. 76, 79).

16. This is echoed in more recent studies (Vogel, 2004).

17. Darwin (1871, p. 86).

18. Langford et al. (2006).

19. Sober and Wilson (1998).

20. Thornton and McAuliffe (2006); Milius (2006).

21. For a more systematic treatment, see Allchin (2009–12).

7. Social Un-Darwinism

1. Bergman (2006).

2. Huxley (1894/1989).

3. Hobbes (1651/1962, p. 118).

4. Malthus (1798/1959, pp. 5–6).

5. Hofstadter (1955, p. 6).

6. Darwin (1859, ch. 3).

7. Spencer (1851/1969, 1852a, 1852b, 1864/1924).

8. Moore (1903).

9. Allchin and Werth (2016).

10. Quoted in Lewontin, Rose, and Kamin (1984, p. 309).

8. A More Fitting Analogy

1. Darwin (1859, p. 81).

2. Darwin (1859, pp. 127, 81).

3. Darwin (1859, p. 5).

4. Darwin (1859, p. 80).

5. Darwin (1859, pp. 90–95).

6. Darwin (1859, pp. 88–89).

7. Darwin (1859, pp. 67, 110).

8. Darwin (1859, p. 490).

9. Gross (1996); Postgate (1994).

10. *Discovery News* (2015); Pennisi (2006).

11. Darwin (1859, p. 489).

9. The Domesticated Gene

1. de Waal (2009, pp. 38–45).

2. Dawkins (2006).

3. Nowak, Tarnita, and Wilson (2010).

4. Brosnan and Bshary (2010).

5. Hamilton (1964); Trivers (1971).

6. Allchin (2009).

7. For a fuller summary, see Allchin (2009–12).

8. de Waal (2009, p. 171).

9. de Waal (1989).

10. Clutton-Brock (2009); Dugatkin (1997).

11. Stephens, McLinn, and Stevens (2002).

12. Woolfenden and Fitzpatrick (1978).

13. J. Smith, Van Dyken, and Zee (2010).

14. Nowak, Tarnita, and Wilson (2010).

15. de Waal (2009, p. 176).

16. Wilkinson (1990).

17. Darwin (1871, p. 86).

18. Boyd, Gintis, and Bowles (2010).

19. Boehm (1999, 2012).

20. Melis and Semmann (2010).

21. Hauert et al. (2007).

22. Rand et al. (2009).

23. Ule et al. (2009).

24. Sigmund (2010).

25. Gächter, Herrmann and Thöni (2010); Henrich et al. (2006).

26. Henrich et al. (2010).

27. Sober and Wilson (1998).

28. Warneken and Tomasello (2006).

29. de Waal (1996, 2009); de Waal and Suchak (2010).

30. de Waal (2006).

31. Darwin (1871).

32. Gazzaniga (2008).

33. de Waal (1996, 2009).

34. Lewontin (1993); Rose (1997).

35. Allchin and Werth (2016).

36. Jensen (2010).

37. Miller (2010).

38. De Dreu et al. (2010).

39. Bowles (2009).

40. Kuhn (1970).

10. A Comedy of Scientific Errors

1. Johnson (2008); Magiels (2010); Nash (1957); Schofield (2004).

2. Priestley (1781, pp. 52–53).

3. Quoted in Nash (1957, p. 360).

4. Franklin (1837).

5. For example, see Matthews (2009).

11. Nobel Ideals and Noble Errors

1. Allchin (1996); Carpenter (2000).

2. Darden (1998); Livio (2013).

3. Magner (2002, pp. 357–359); Nye (2007).

4. Pauling (1970).

5. Hurd (2007).

6. Allchin (2007a).

7. Bibel (1988); Magner (2002, pp. 278–285); Silverstein (1989).

8. Crick (1958).

9. See http://nobelprize.org/nobel_prizes/medicine/laureates/2006/illpres/2_central_dogma.html.

10. Judson (1979, p. 337).

11. Crick (1970).

12. Crick (1981).

13. Shepherd (2007).

14. Eccles (1952, pp. 271–286).

15. Eccles (1989, pp. xiii, 236–245).

16. Allchin (2002); Prebble and Weber (2003).

17. Longino (1990); Solomon (2001).

18. Brush (1974).

12. Celebrating Darwin's Errors

1. Ghiselin (1969).

2. Dennison (2006).

3. Eiseley (1961, pp. 216–221); Ghiselin (1969, pp. 162–164).

4. Quoted in Eiseley (1961, p. 191).

5. Sulloway (1982); Browne (1996, pp. 359–360).

6. Eriksson et al. (2008).

7. Gould (1980).

8. Browne (1996, pp. 316–319); Ghiselin (1969, pp. 21–30).

9. Browne (1996, pp. 376–378, 431–433); Rudwick (1974).

10. Darwin (1845, p. 218).

11. Browne (1996, pp. 234–253, 382–383); Herbert (1974, 1977).

12. Eiseley (1961, pp. 303–314, quotes on p. 303).

13. Katz (2000).

14. Barkan (1992); Stepan (1982).

15. Browne (1996, pp. 196–199, 213–214, 244–246).

16. Browne (1996, pp. 542–543); Ghiselin (1969, pp. 48–49, 59–61); Young (1975).

17. Echoed in Darwin (1859, p. 67).

18. Mayr (1994).

19. Hsü (1986).

13. Science beyond Scientists

1. Griswold (2011).

2. Agin (2006).

3. Goldacre (2010, 2014).

4. Park (2000).

5. Oreskes and Conway (2010). See also Kenner (2015).

6. Michaels (2008).

7. Mooney (2005a).

8. McGarity and Wagner (2008).

9. Huber (1991).

10. See https://www.aaas.org/page/national-conference-lawyers-and-scientists.

11. B. Martin (1991); Rampton and Stauber (2001); Toumey (1997).

12. Gardner (1957, 1981).

13. For example, Agin (2006); Park (2000); Pigliucci (2010); Shermer (2002).

14. Duyff (2002).

15. Park (2003).

16. "Untangling media messages and public policies" (2012).

17. Freedman (2010, p. 75).

18. Kahneman (2011); Lehrer (2009).

19. Gilovich (1991); Hallinan (2009); Sutherland (1992).

20. Shermer (2002, pp. 279–313).

21. Freedman (2010, pp. 76–80, 184, 217–224).

22. Gilovich (1991); Hallinan (2009); Kahneman (2011); Lehrer (2009); Sutherland (1992).

23. Brownlee and Lenzer (2011); Goodnough (2011); Harris (2011); Rosenberg (2011).

24. Hardwig (1991).

25. Gaon and Norris (2001).

14. Skepticism and the Architecture of Trust

1. Brownlee and Lenzer (2011); Harris (2011).

2. Goodnough (2011); Rosenberg (2011).

3. Allchin (2015).

4. Shapin (1994).

5. Hardwig (1991).

6. Latour and Woolgar (1979, pp. 131–135).

7. Collins and Evans (2007).

8. Goldman (2001).

9. Shay and Pinch (2005).

10. Toumey (1997, pp. 81–95).

11. Park (2000, pp. 98–106).

12. Goldacre (2010, pp. 112–130).

13. Kahneman (2011).

14. Oreskes and Conway (2010).

15. Spence et al. (1993); Toumey (1997, pp. 3–4).

16. Rampton and Stauber (2001, pp. 276–278).

17. Epstein (1995).

18. Epstein (1996, p. 338).

19. González (2001).

20. Bass (1990, pp. 1–50).

21. Allchin (1999).

22. Park (2000, pp. 140–161).

15. Science Con Artists

1. Toumey (1997, p. 6).

2. Goldacre (2010, pp. 131–146).

3. Oreskes and Conway (2010).

4. Mooney (2005b).

5. Goldman (2001); McGarity and Wagner (2008); Michaels (2008); Oreskes and Conway (2010).

6. Freedman (2010); Kahneman (2011); Rampton and Stauber (2001, pp. 291–294).

7. Yahya (2006).

8. American Petroleum Institute (2012, "About").

9. Center for Media and Democracy (2012); Rampton and Stauber (2001).

10. Michaels (2008, pp. 53–55); Oreskes and Conway (2010, pp. 244–245).

11. McGarity and Wagner (2008, pp. 76–79); Rampton and Stauber (2001, pp. 200–201).

12. Freedman (2010, p.66).

13. D. M. Cook et al. (2007).

14. Milloy (2001).

15. D. Murray et al. (2001, quotes on pp. 131, 159, 193).

16. Goldacre (2010).

17. B. Martin (1991); Toumey (1997, pp. 63–80).

18. Toumey (1997, pp. 81–95).

19. Miller (2011).

20. De Dreu et al. (2010).

21. Rampton and Stauber (2001, pp. 251, 291–294).

22. Kenner (2015); McGarity and Wagner (2008, pp. 146–149); Michaels (2008); Oreskes and Conway (2010).

23. Quoted in Michaels (2008, pp. x, 11).

24. Michaels (2008, pp. ix, xi).

25. McGarity and Wagner (2008, pp. 204–228); Rampton and Stauber (2001, p. 294).

26. Oreskes and Conway (2010, pp. 190–197).

27. Kahneman (2011).

28. Oreskes and Conway (2010, p. 203).

29. http://heartland.org/publications (accessed April 1, 2013).

30. Oreskes and Conway (2010, pp. 208–211).

16. Male, Female, and/or —?

1. Eugenides (2002).

2. Online Mendelian Inheritance in Man (2012, No. 607306).

3. Bainbridge (2003, pp. 39–56); Berec, Schembri, and Boukal (2005); Roughgarden (2004, pp. 203–205).

4. Bainbridge (2003, pp. 17–21, 58); Roughgarden (2004, pp. 198–202).

5. Dreger (1998, pp. 37–39, 84–106).

6. Roughgarden (2004, pp. 35–42).

7. Roughgarden (2004, pp. 30–35).

8. Roughgarden (2004, pp. 30–31).

9. Dreger (1998, pp. 37, 73–74, 147–149, 159–161); Fausto-Sterling (2000, pp. 51, 53); Roughgarden (2004, p. 41).

10. Dreger (1998, pp. 7–8); Fausto-Sterling (2000, pp. 170–194); Roughgarden (2004, pp. 215–221).

11. Crews (1987).

12. Demir and Dickson (2005); Miller (2005).

13. Mansfield (2006).

14. Roughgarden (2004, pp. 45–48).

15. On sex-biased behaviors, see Fausto-Sterling (2000, pp. 115–232).

16. Fausto-Sterling (2000, p. 68 [after Milton Diamond]).

17. Bell (1982); Hoekstra (1987).

18. Whitfield (2004).

19. Darwin (1877).

20. Darwin (1887/1958, p. 134).

21. Darwin (1877, p. 258).

22. Roughgarden (2004, pp. 24–25, 75–93).

23. Parker (2004).

24. Blackless et al. (2000); Dreger (1998, pp. 40–43); Fausto-Sterling (2000, pp. 51–53).

25. Allchin and Werth (2016).

17. Monsters and the Tyranny of Normality

1. Blumberg (2009).

2. Allchin (1996); Gabel and Allchin (2017); Howe (2007).

3. Erard (2007).

4. Garber (2007).

5. Freedberg (2002).

6. Hunter (1784, p. 4).

7. Daston and Park (2001, pp. 204–205).

8. Daston and Park (2001, p. 204); New York Academy of Medicine (2007).

9. Geoffroy Saint-Hilaire (1822/1968, pp. 106, 105, 15).

10. Appel (1987).

11. Howell and Ford (1980, p. 189).

12. Dreger (1998).

13. Allchin and Werth (2016).

18. To Be Human

1. Culotta (2005); Lemonick and Dorfman (2006).

2. Gibbons (2010).

3. *Science* (October 2, 2009).

4. *Science* (April 9, 2010).

5. Potts and Sloan (2010a); Tattersall and DeSalle (2007).

6. DeSalle and Tattersall (2008); Gazzaniga (2008); Potts and Sloan (2010b); Röska-Hardy and Neumann-Held (2008); Shubin (2008); Taylor (2009); Walter (2006); Zimmer (2007).

7. Chedd at al. (2010); Discovery Channel (2009); Rubin (2008); Townsley (2009).

8. Darwin (1871, I: pp. 51–53, 138–143); Moon (1921, pp. 336–339).

9. For example, Dobzhansky (1962, p. 194); H. W. Smith (1961, p. 178); Tattersall (1998, pp. 49–57, 126–134).

10. Dobzhansky (1962, p. 194); Linden (2003, pp. 91–108); Mason (1972).

11. Darwin (1871, I: p. 52).

12. Dobzhansky (1962, p. 194).

13. Gibbons (2007); McGrew (2010).

14. Boesch (1991).

15. Mason (1972, p. 388).

16. Mulcahy and Call (2006).

17. Darwin (1871, I: pp. 53–62); Diamond (1992, pp. 137–167); Dobzhansky (1962, pp. 208–210); Simpson (1967, pp. 287–288); Tattersall (1998, pp. 58–68).

18. Hurford (2004, p. 551).

19. For example, Gazzaniga (2008, pp. 54–66); Potts and Sloan (2010a); Tattersall (1998, pp. 226–233); Tattersall and DeSalle (2007).

20. Jones (1994).

21. Dobzhansky (1962, pp. 213–214); Linden (2003, pp. 57–67); Masson and McCarthy, 1995, pp. 124–132).

22. Panskepp (2005).

23. Dobzhansky (1962, pp. 214–218); Gazzaniga (2008, pp. 203–245); Tattersall (1998, pp. 14–28); but compare with George (1995); Lenain (1997); Linden (2003, pp. 167–176); Masson and McCarthy (1995, pp. 192–211).

24. de Waal (2009).

25. Darwin (1871, ch. 3); Gazzaniga (2008, pp. 113–157).

26. Allchin (2009–2012).

27. Gazzaniga (2008, pp. 7–37).

28. Despain (2010).

29. Darwin (1872/1965, p. 309).

30. Twain (1897, I: ch. 27).

31. Tattersall and DeSalle (2007).

32. Potts and Sloan (2010a); see also Gazzaniga (2008, pp. 79–112).

33. E. Morgan (1990).

34. Walter (2006).

35. Gazzaniga (2008, p. 45); Shubin (2008, p. 189).

36. Culotta (2005, p. 1468).

37. Walter (2006, pp. 143–161).

38. Tattersall (1998, p. 13).

39. Gazzaniga (2008, pp. 44, 45, 326).

40. Allchin and Werth (2016).

41. Culotta (2005).

42. Taylor (2009).

43. Linden (2003); Masson and McCarthy (1995).

44. Shubin (2008).

19. Genes R Us

1. *Life Magazine* (1996), R. Murray (2012).

20. The Peppered Moths: A Study in Black and White

1. Kettlewell (1973, plate 9.1).
2. Kettlewell (1955; 1956; 1973, pp. 134–136).
3. Kettlewell (1959, p. 51).
4. Kettlewell (1959, p. 51).
5. Kettlewell (1973, pp. 106–107).
6. Endler (1986); Kettlewell (1959).
7. Hagen (1993, 1999); Hagen, Allchin, and Singer (1996, pp. 1–10).
8. Rudge (1999).
9. B. Martin (1991); Toumey (1997, pp. 63–80).
10. Toumey (1997).
11. For recent accounts updating the science of the peppered moth, see L. M. Cook et al. (2012); Majerus (1998, 2009).

21. Alexander Fleming's "Eureka" Moment

1. Ho (1999).
2. Maurois (1959, p. 206); see also Fleming (1945/1964, p. 1).
3. Macfarlane (1985) informs the text throughout.
4. Fleming (1929, p. 226).
5. Judson (1981); see also list in the Wikipedia article "History of Penicillin."
6. Macfarlane (1985, pp. 177–180, 187–189).
7. Ho (1999).
8. Allchin (2013, pp. 46–76).
9. Allchin and Werth (2016).

22. Round versus Wrinkled: Gregor Mendel as Icon

1. Online Mendelian Inheritance in Man (2012).
2. Rodgers (1991).
3. Guilfoile (1997); Reid and Ross (2011).
4. Darbishire (1908).
5. Bateson (1909, pp. 28–29, 53).
6. Bateson (1902, p. 152).
7. Bateson (1902, p. 129).
8. Mendel (1866, §4).
9. Mendel (1866, §4).
10. Mendel (1866, §4).
11. Mendel (1866, §8).
12. Mendel (1866, §10).
13. Bateson (1909, p. 53).
14. Tschermak (1900/1950, p. 44).
15. Correns (1900/1950, p. 34).
16. Allchin (2005).
17. Di Trocchio (1991, p. 493); Orel (1984, p. 44).
18. Mendel (1866, §8).

19. Mendel (1866, §2).

20. Mendel (1866, §8).

21. Bateson (1909, p. 29).

22. Wang (2002).

23. Coyne (2002).

24. Mendel (1866, §3).

25. Bateson (1909, p. 28).

26. Endersby (2007); Mendel (1869/1966); Nogler (2006).

27. Sapp (1990).

23. William Harvey and Capillaries

1. Elkana and Goodfield (1968).

2. Harvey (1628/1952, ch. 7, p. 283).

3. Harvey (1649a/1952, pp. 308–309).

4. Harvey (1649b/1952, p. 322).

5. Harvey (1649a/1952, p. 311).

6. Harvey (1649a/1952, p. 311).

7. Malpighi (1661/1929, p. 8).

8. Adelmann (1966, I: pp. 171–198).

9. Daintith and Gjertsen (2003); Lawson (2000); Leinhard (1997b); Lewis (1988); Phillips (2004); Starr (2002). The error appeared in the Wikipedia entry ("William Harvey," 2015) until it was corrected, on September 6, 2015.

10. Elkana and Goodfield (1968).

11. Lawson (2000); Lewis (1988).

12. Allchin (2004).

24. The Tragic Hero of Childbed Fever

1. Leinhard (1997a).

2. Colyer (2000).

3. Semmelweis Society International (2009).

4. "Ignaz Semmelweis" (n.d.).

5. Colyer (2000).

6. Horton (2004).

7. Hanninen, Farago, and Monos (1983); Leinhard (1997a).

8. Carter (1983).

9. Johnston (1972, p. 224).

10. Carter (1983, p. 41).

11. Colyer (2000).

12. Carter (1983, p. 57, transl. by S. B. Nuland and Gyorgyey).

13. Carter (1983, p. 53).

25. Respect for Life

1. The phrase "grandeur in this view of life" to describe the tree of life comes directly from Charles Darwin, in the closing of his epic *Origin of Species*.

2. http://www.statista.com/statistics/253904/us-dog-food-production-by-category (accessed February 20, 2016).

3. http://www.statista.com/statistics/253983/pet-market-sales-in-the-us-by-category (accessed February 20, 2016).

26. Hands-Off Dissection?

1. National Association of Biology Teachers (2008).
2. Russell (1996, p. 2).
3. Russell (1996, p. 6).
4. Cross and Cross (2004).
5. Mathé, J. (1978, p. 22).
6. French (1999).

27. Organisms, Modified, Genetically

1. Clark (2014).
2. Guilhem (2013).
3. Genetic Science Learning Center (2014).
4. Clark (2014).
5. Hallinan (2009); Lehrer (2009).
6. Clark (2014).
7. Konnikova (2013).
8. Harmon (2014).
9. Diamond (1987).

28. Close to Nature

1. Carroll (2009); Flannery (1991); Shubin (2008).

Epilogue: Challenging Sacred Bovines, Fostering Creativity

1. Burdick Group (1982); de Bono (1970).
2. Barron and Eisner (1982); Burdick Group (1982); de Bono (1970); Jackson and Mendoza (1979); Kelley (2001); Ness (2012); Ortony (1993).
3. Burdick Group (1982); May (1975).
4. Burke (1978); Perkins (1981).
5. Kuhn (1970); Thagard (1992).

Afterword for Educators: Sacred Bovines in the Classroom

1. For more on teaching the nature of science, see Allchin (2013).
2. NGSS Lead States (2013).
3. Allchin, Andersen, and Nielsen (2014).
4. Allchin (2012). For sample cases, see http://ships.umn.edu/modules and http://doing-biology.net.
5. NGSS Lead States (2013).

REFERENCES

Adelmann, H. B. (1966). *Marcello Malpighi and the Evolution of Embryology*. Ithaca, NY: Cornell University Press.

Agin, D. (2006). *Junk Science*. New York: St. Martin's Press.

Allchin, D. (1996). Christian Eijkmann & the cause of beriberi. In J. Hagen, D. Allchin, & F. Singer, *Doing Biology* (pp. 116–127). Glenview, IL: Harper Collins.

Allchin, D. (1999). Do we see through a social microscope? Credibility as a vicarious selector. *Philosophy of Science, 60* (Proceedings), S287–S298.

Allchin, D. (2002). To err and win a Nobel Prize: Paul Boyer, ATP synthase and the emergence of bioenergetics. *Journal of the History of Biology, 35,* 149–172.

Allchin, D. (2004). Pseudohistory and pseudoscience. *Science & Education, 13,* 179–195.

Allchin, D. (2005). The dilemma of dominance. *Biology and Philosophy, 20,* 427–451.

Allchin, D. (2007a). Albert Szent-Gyorgyi. In Noretta Koertge, *New Dictionary of Scientific Biography* (Vol. 24, pp. 567–573). Detroit, MI: Charles Scribner's Sons.

Allchin, D. (2007b). Alex Benjamin Novikoff. In Noretta Koertge, *New Dictionary of Scientific Biography* (Vol. 23, pp. 281–285). Detroit, MI: Charles Scribner's Sons.

Allchin, D. (2009). Teaching the evolution of morality: Status and resources. *Evolution: Education and Outreach, 2,* 629–635.

Allchin, D. (2009–12). *The Evolution of Morality* [website]. Available at http://evolutionofmorality.net

Allchin, D. (2012). The Minnesota Case Study Collection: New historical inquiry cases for nature of science education. *Science & Education, 21,* 1263–1282.

Allchin, D. (2013). *Teaching the Nature of Science: Perspectives and Resources*. St. Paul, MN: SHiPS Press.

Allchin, D. (2015). Global warming: Scam, fraud, or hoax? *American Biology Teacher, 77,* 308–312.

Allchin, D., Andersen, H. M., & Nielsen, K. (2014). Complementary approaches to teaching nature of science: Integrating student inquiry, contemporary cases and historical cases in classroom practice. *Science Education, 98,* 461–486.

Allchin, D., & Werth. A. (2016). The naturalizing error. *Journal for the General Philosophy of Science*. doi:10.1007/s10838-016-9336-x

Allen, G. E. (1978). *Thomas Hunt Morgan: The Man and His Science*. Princeton, NJ: Princeton University Press.

American Petroleum Institute. (2012). Energy Answered. Washington, DC. Available at https://www.depts.ttu.edu/vpr/uptec/pro-hf.php

Anand, P. (2002). Decision-making when science is ambiguous. *Science, 295,* 1839.

Appel, T. (1987). *The Cuvier-Geoffroy Debate*. New York: Oxford University Press.

Bainbridge, D. (2003). *The X in Sex*. Cambridge, MA: Harvard University Press.

Barkan, E. (1992). *The Retreat of Scientific Racism*. Cambridge, UK: Cambridge University Press.

Barrett, P., Gautrey, P. J., Herbert, S., Kohn D., & Smith, S. (1987). *Charles Darwin's Notebooks (1836–1844)*. Ithaca, NY: Cornell University Press.

Barron, F., & Eisner, E. (1982). *Creativity, The Human Resource* [audio filmstrip]. San Francisco: Burdick Group and Chevron/Standard Oil of California.

Bartal, I. B., Decety, J., & Mason, P. (2011). Helping a cagemate in need: Empathy and prosocial behavior in rats. *Science, 334,* 1427–1430.

Bass, T. (1990). *Camping with the Prince and Other Tales of Science in Africa*. Boston: Houghton Mifflin.

Bateson, W. (1902). The facts of heredity in the light of Mendel's discovery. *Reports to the Evolution Committee of the Royal Society, London, 1,* 125–160.

Bateson, W. (1909). *Mendel's Principles of Heredity*. Cambridge, UK: Cambridge University Press.

Bauer, H. H. (1992). *Scientific Literacy and the Myth of the Scientific Method*. Urbana: University of Illinois Press.

Bazerman, C. (1988). *Shaping Written Knowledge: The Genre and Activity of the Experimental Article in Science*. Madison: University of Wisconsin Press.

Bell, G. (1982). *The Masterpiece of Nature*. Berkeley: University of California Press.

Berec, L., Schembri, P. J., & Boukal, D. S. (2005). Sex determination in *Bonellia viridis* (Echiura: Bonelliidae): Population dynamics and evolution. *Oikos, 108,* 473–484.

Bergman, J. (2006). Darwin's influence on ruthless laissez faire capitalism. Dallas, TX: Institute for Creation Research. Available at http://www.icr.org/article/darwins-influence-ruthless-laissez-faire-capitalis/ (accessed December 27, 2016).

Bibel, D. J. (1988). *Milestones in Immunology: A Historical Exploration*. Madison, WI: Science Tech Publishers.

Blackless, M., Charuvastra, A., Derryck, A., Fausto-Sterling, A., Lauzanne, K., & Lee, E. (2000). How sexually dimorphic are we? Review and synthesis. *American Journal of Human Biology, 12,* 151–166.

Blumberg, M. S. (2009). *Freaks of Nature and What They Tell Us about Development and Evolution*. Oxford: Oxford University Press.

Boehm, C. (1999). *Hierarchy in the Forest: The Evolution of Egalitarian Behavior*. Cambridge, MA: Harvard University Press.

Boehm, C. (2012). *Moral Origins: The Evolution of Virtue, Altruism, and Shame*. New York: Basic Books.

Boesch, C. (1991). Teaching in wild chimpanzees. *Animal Behaviour, 41,* 530–532.

Bondison, J. (1997). *A Cabinet of Medical Curiosities*. Ithaca, NY: Cornell University Press.

Bondison, J. (2004). *The Two-Headed Boy, and Other Medical Marvels*. Ithaca, NY: Cornell University Press.

Bowles, S. (2009). Did warfare among ancestral hunter-gatherers affect the evolution of human social behaviors? *Science, 324,* 1293–1298.

Boyd, R., Gintis, H., & Bowles, S. (2010). Coordinated punishment of defectors sustains cooperation and can proliferate when rare. *Science, 328,* 617–620.

Bradie, M. (1994). *The Secret Chain*. Albany: State University of New York Press.

Brosnan, S. F., & Bshary, R. (2010). Cooperation and deception: From evolution to mechanisms. *Philosophical Transactions of the Royal Society B, 365,* 2593–2598.

Browne, J. (1996). *Charles Darwin: Voyaging*. Princeton, NJ: Princeton University Press.

Browne, J. (2003). *Charles Darwin: The Power of Place*. Princeton, NJ: Princeton University Press.

Brownlee, S., & Lenzer, J. (2011, October 9). The bitter fight over prostate screening—and why it might be better not to know. *New York Times Magazine*, pp. 40–43, 57.

Brush, S. (1974). Should the history of science be rated 'X'? *Science, 18,* 1164–1172.

Burdick Group (1982). *Creativity: The Human Resource* [Exhibit]. San Francisco: Standard Oil Company of California.

Burke, J. (1978). *Connections*. Boston: Little, Brown.

Campbell, D. (1974). Evolutionary epistemology. In P. A. Schilpp (Ed.), *The Philosophy of Karl Popper* (pp. 413–463). La Salle, IL: Open Court.

Carpenter, K. J. (2000). *Beriberi, White Rice and Vitamin B*. Berkeley: University of California Press.

Carroll, S. B. (2009). *Remarkable Creatures: Epic Adventures in the Search for the Origins of Species*. New York: Mariner Books.

Carter, K. C. (1983). Translator's introduction. In I. Semmelweis (1861/1983), *The Etiology, Concept, and Prophylaxis of Childbed Fever* (pp. 3–58). Madison: University of Wisconsin Press.

Center for Media and Democracy. (2012). Foundation for Clean Air Progress. Madison,WI: Author. Available at http://www.sourcewatch.org/index.php?title=Foundation_for_Clean_Air_Progress

Chedd, G., Buckner, R., Dunbar, R., & Saxe, R. (Producers). (2010). *The Human Spark* [TV program] (3 Pts.). Public Broadcasting System, Chedd-Angier-Lewis Productions, and THIRTEEN in association with WNET.org.

Clark, R. (2014, January 1914). GMO salmon [Web log post]. Mr. Clark's AP Biology Blog. Available at http://rodneylynnclark.blogspot.com/2014/01/5th-post-thursday-1914-gmo-salmon-final.html

Clutton-Brock, T. (2009). Cooperation between non-kin in animal societies. *Nature, 462,* 51–57.

Clutton-Brock, T. H., & Parker, G. A. (1995). Punishment in animal societies. *Nature, 37,* 209–216.

Collins, H., & Evans, R. (2007). *Rethinking Expertise*. Chicago: University of Chicago Press.

Colyer, C. (2000). Death in a Viennese maternity ward. *Journal of College Science Teaching, 29,* 297–300. Reprinted in C. E. Herreid, N. A. Schiller, & K. F. Herreid (Eds.) (2012), *Science Stories: Using Case Studies to Teach Critical Thinking* (pp. 39–44). Arlington, VA: NSTA Press. Also available at http://sciencecases.lib.buffalo.edu/cs/collection/detail.asp?case_id=429&id=429 (posted December 8, 1999; accessed September 2, 2013).

Cook, D. M., Boyd, E. A., Grossmann, C., & Bero, L. A. (2007). Reporting science and conflicts of interest in the lay press. *PLoS ONE, 2*(12), e1266. doi:10.1371/journal.pone.0001266

Cook, L. M., Grant, B. S., Saccheri, I. J., & Mallet, J. (2012). Selective bird predation on the peppered moth: The last experiment of Michael Majerus. *Biology Letters, 8*(4), 609–612. doi:10.1098/rsbl.2011.1136

Cooper, J., & Carlsmith, K. M. (2001). Cognitive dissonance. In N.J. Smelzer, P.B. Balte, *International Encyclopedia of the Social and Behavioral Sciences* (pp. 2112–2114). Amsterdam: Elsevier.

Correns, C. (1900/1950). G. Mendel's law concerning the behavior of progeny of varietal hybrids [Eng. trans.]. *Genetics, 35*(5, 2), 33–41.

Coyne, C. (Curator). (2002). *Pisum*. The G. A. Marx Pea Genetic Stock Center. Agricultural Research Service. Available at https://www.ars.usda.gov/pacific-west-area/pullman-wa/plant-germplasm-introduction-and-testing-research/docs/ga-marx-pea-genetic-stock-collection/main/ (accessed December 28, 2016).

Crews, D. (1987, December), Courtship in unisexual lizards: A model for brain evolution. *Scientific American, 225,* 116–121.

Crick, F. H. C. (1958). On protein synthesis. *Symposia of the Society for Experimental Biology, 12,* 138–163.

Crick, F. (1970). Central dogma of molecular biology. *Nature, 227,* 561–563.

Crick, F. (1981). *Life Itself.* New York: Simon & Schuster.

Cross, T. R., & Cross, V. E. (2004). Scalpel or mouse? A statistical comparison of real and virtual frog dissections. *American Biology Teacher, 66,* 408–411.

Culotta, E. (2005). Chimp genome catalogs differences with humans. *Science, 309,* 1468–1469.

Daintith, J., & Gjertsen, D. (Eds.). (2003). William Harvey. *A Dictionary of Scientists.* Oxford and New York: Oxford University Press.

Darbishire, A. D. (1908). On the result of crossing round with wrinkled peas, with especial reference to their starch grains. *Proceedings of the Royal Society, 80*(8), 122–135. Available at http://rspb.royalsocietypublishing.org/content/80/537/122.full.pdf+html (accessed March 23, 2012).

Darden, L. (1998). The nature of scientific inquiry. Available at http://www.philosophy.umd.edu/Faculty/LDarden/sciinq/

Darwin, C. R. (1845). *Journal of Researches* (2nd ed.). London: John Murray.

Darwin, C. (1859). *On the Origin of Species.* London: John Murray.

Darwin, C. (1871). *The Descent of Man.* London: John Murray.

Darwin, C. (1872/1965). *The Expression of the Emotions in Man and Animals.* Chicago: University of Chicago Press.

Darwin, C. (1877). *The Different Forms of Flowers on Plants of the Same Species.* London: John Murray.

Darwin, C. (1887/1958). *The Autobiography of Charles Darwin.* New York: W. W. Norton.

Daston, L., & Park, K. (2001). *Wonders and the Order of Nature.* New York: Zone Books.

Dawkins, R. (2006). Introduction. *The Selfish Gene* (30th anniv. ed.). Oxford: Oxford University Press.

de Bono, E. (1970). *Lateral Thinking: Creativity Step by Step.* New York: Harper & Row.

De Dreu, C. K. W., Greer, L. L., Handgraaf, M. J. J., Shalvi, S., Van Kleef, G. A., Baas, M., Velden, F. S., Van Dijk, E., & Feith, S. W. W. (2010). The neuropeptide oxytocin regulates parochial altruism in intergroup conflict among humans. *Science, 328,* 1408–1411.

Delbrück, M. (1954). On the replication of desoxyribonucleic acid (DNA). *Proceedings of the National Academy of Sciences of the USA, 40,* 783–788.

Della Porta, J. B. (1658). *Natural Magick.* Available at https://archive.org/details/naturalmagick00port

Demir, E. & Dickson, B. J. (2005). *Fruitless* splicing specifies male courtship behavior in *Drosophila. Cell, 121,* 785–794.

Dennison, R. (2006). Charles Darwin: Biology's most important experimenter. In W. McComas (Ed.), *Investigating Evolutionary Biology in the Laboratory* (pp. 96–106). Dubuque, IA: Kendall Hunt.

DeSalle, R., & Tattersall, I. (2008). *Human Origins: What Bones and Genomes Tell Us about Ourselves*. College Station: Texas A&M University Press.

Despain, D. (2010, February 26). Cross-discipline effort tracks evolution of human uniqueness and modern behavior. *Scientific American*. Available at http://www.scientificamerican.com/article.cfm?id=human-uniqueness-anthropology

de Waal, F. (1989). Food sharing and reciprocal obligations among chimpanzees. *Journal of Human Evolution, 18,* 433–459.

de Waal, F. (1996). *Good Natured: The Origins of Right and Wrong in Humans and Other Animals*. Cambridge, MA: Harvard University Press.

de Waal, F. (2006). *Primates and Philosophers: How Morality Evolved*. Princeton, NJ: Princeton University Press.

de Waal, F. (2009). *The Age of Empathy: Nature's Lessons for a Kinder Society*. New York: Harmony Books.

de Waal, F. B. M., & Suchak, M. (2010). Prosocial primates: Selfish and unselfish motivations. *Philosophical Transactions of the Royal Society B, 365,* 2711–2722.

Diamond, J. (1987, May). The worst mistake in the history of the human race. *Discover Magazine,* 64–66. Available at http://discovermagazine.com/1987/may/02-the-worst-mistake-in-the-history-of-the-human-race

Diamond, J. (1992). *The Third Chimpanzee*. New York: Harper Collins.

Discovery Channel. (2009). *Discovering Ardi* [TV show]. Silver Spring, MD: Discovery Communications.

Discovery News. (2015). Sea slug uses gene from algae to live like a plant. Available at http://news.discovery.com/animals/sea-slug-uses-gene-from-algae-to-live-like-a-plant-150204.htm

Di Trocchio, F. (1991). Mendel's experiments: A reinterpretation. *Journal of the History of Biology, 24,* 485–519.

DNA Learning Center. (2011). A half DNA ladder is a template for copying the whole. *DNA from the Beginning*, Concept 20. Cold Spring Harbor, NY: Cold Spring Harbor Laboratory. Available at http://www.dnaftb.org/20/animation.html

Dobzhansky, T. (1962). *Mankind Evolving*. New Haven, CT: Yale University Press.

Dreger, A. D. (1998). *Hermaphrodites and the Medical Invention of Sex*. Cambridge, MA: Harvard University Press.

Dugatkin, L. A. (1997). *Cooperation among Animals: An Evolutionary Perspective*. Oxford: Oxford University Press.

Duyff, R. (2002). *American Dietetic Association Complete Food and Nutrition Guide* (2nd ed.). New York: John Wiley.

Eccles, J. C. (1952). *The Neurophysiological Basis of Mind*. Oxford: Oxford University Press.

Eccles, J. C. (1989). *Evolution of the Brain: Creation of the Self*. London: Routledge.

Einstein, A. (1954). *Ideas and Opinions*. New York: Dell.

Eiseley, L. (1961). *Darwin's Century*. Garden City, NY: Anchor Books.

Elkana, Y., & Goodfield, J. (1968). Harvey and the problem of the "capillaries." *Isis, 59,* 61–73.

Endersby, J. (2007). *A Guinea Pig's History of Biology*. Cambridge, MA: Harvard University Press.

Endler, J. (1986). *Natural Selection in the Wild*. Princeton, NJ: Princeton University Press.

Epstein, S. (1995). The construction of lay expertise: AIDS activism and the forging of credibility in the reform of clinical trials. *Science, Technology, & Human Values, 20,* 408–437.

Epstein, S. (1996). *Impure Science: AIDS, Activism and the Politics of Knowledge*. Berkeley: University of California Press.

Erard, M. (2007). Read my slips: Speech errors show how language is processed. *Science, 317*, 1674–1676.

Eriksson, J., Larson, G., Gunnarsson, U., Bed'hom, B., Tixier-Boichard, M., et al. (2008). Identification of the yellow skin gene reveals a hybrid origin of the domestic chicken. *PLoS Genetics, 4*(2), e1000010. doi:10.1371/journal.pgen.1000010

Eugenides, J. (2002). *Middlesex*. New York: Picador.

Farber, P. (1994). *The Temptations of Evolutionary Ethics*. Berkeley: University of California Press.

Fausto-Sterling, A. (2000). *Sexing the Body*. New York: Basic Books.

Findlen, P. (2002). Inventing nature. In Smith, P. H., & Findlen, P. (Eds.), *Merchants and Marvel*, (pp. 297–323). New York: Routledge.

Flannery, M. (1991). *Bitten by the Biology Bug*. Reston, VA: National Association of Biology Teachers.

Fleming, A. (1929). On the antibacterial action of cultures of a *Penicillium*, with special reference to their use in the isolation of *B. influenzae. British Journal of Experimental Pathology, 10*, 226–236.

Fleming, A. (1945/1964). Penicillin. In *Nobel Lectures, Physiology or Medicine 1942–1962*. Amsterdam: Elsevier. Available at http://www.nobelprize.org/nobel_prizes/medicine/laureates/1945/fleming-lecture.pdf

Franklin, B. (1837). *Animal Magnetism. Report of Dr. Benjamin Franklin, and Other Commissioners, Charged by the King of France, with the Examination of the Animal Magnetism, as Now Practised in Paris*. Philadelphia: H. Perkins.

Freedberg, D. (2002). *The Eye of the Lynx*. Chicago: University of Chicago Press.

Freedman, D. H. (2010). *Wrong: Why Experts Keep Failing Us—and How to Know When Not to Trust Them*. New York: Little, Brown.

French, R. (1999). *Dissection and Vivisection in the European Renaissance*. Aldershot, UK: Ashgate.

Gabel, K. & Allchin, D. (2017). Archibald Garrod and the Black Urine Disease. St. Paul, MN: SHiPS Resource Center. Available at http://shipseducation.net/modules/biol/garrod.htm

Gächter, S., Herrmann, B., & Thöni, C. (2010). Culture and cooperation. *Philosophical Transactions of the Royal Society B, 365*, 2651–2661.

Ganguli, I. (2006, June 30). Mice show evidence of empathy. *The Scientist*. Available at http://www.the-scientist.com/news/display/23764/

Gaon, S., & Norris, S. P. (2001). The undecidable grounds of scientific expertise: Science education and the limits of intellectual independence. *Journal of Philosophy of Education, 35*, 187–201.

Garber, K. (2007). Autism's cause may result in abnormalities at the synapse. *Science, 317*, 190–191.

Gardner, M. (1957). *Fads and Fallacies in the Name of Science*. Mineola, NY: Dover.

Gardner, M. (1981). *Science: Good, Bad, and Bogus*. Buffalo, NY: Prometheus Books.

Gazzaniga, M. (2008). *Human: The Science behind What Makes Your Brain Unique*. New York: Harper Collins.

Genetic Science Learning Center. (2014). Evolution of corn. Salt Lake City: University of Utah. Available at http://learn.genetics.utah.edu/content/selection/corn

Geoffroy Saint-Hilaire, E. (1822/1968). *Philosophie Anatomique. II. Des Monstruosités Humaines*. Brussels: Culture et Civilisation Impression Anastalique.

George, D. (1995). *Ruby, the Painting Pachyderm of the Phoenix Zoo*. New York: Delacorte Press.

Ghiselin, M. (1969). *The Triumph of the Darwinian Method*. Chicago: University of Chicago Press.

Gibbons, A. (2007). Spear-wielding chimps seen hunting bush babies. *Science, 315*, 1063.

Gibbons, A. (2010). Close encounters of the prehistoric kind. *Science, 328*, 680–684.

Gilovich, T. (1991). *How We Know What Isn't So*. New York: Free Press.

Goldacre, B. (2010). *Bad Science: Quacks, Hacks and Big Pharma Flacks*. New York: Faber and Faber.

Goldacre, B. (2014). *I Think You'll Find It's a Bit More Complicated than That*. London: Fourth Estate.

Goldman, A. I. (2001). Experts: Which ones should you trust? *Philosophy and Phenomenological Research, 63*, 85–110.

González, R. J. (2001). *Zapotec Science: Farming and Food in the Northern Sierra of Oaxaca*. Austin: University of Texas Press.

Goodnough, A. (2011, December 10). Scientists say cod are scant; nets say otherwise. *New York Times*, pp. A20, A27.

Gould, S. J. (1980). *Bathybius* and *Eozoon*. In *The Panda's Thumb* (pp. 236–244). New York: W. W. Norton.

Grant, P. R., & Grant, B. R. (2002). Unpredictable evolution in a 30-year study of Darwin's finches. *Science, 296*, 707–711.

Gregory, R. P. (1903). The seed characters of *Pisum sativum. The New Phytologist, 2*(10), 226–228. http://www.jstor.org/stable/2427335. Also available at http://onlinelibrary. wiley.com/doi/10.1111/j.1469-8137.1903.tb07339.x/pdf (accessed March 24, 2012).

Griswold, E. (2011, November 17). The fracturing of Pennsylvania: situation normal all fracked up. *New York Times Magazine*, pp. 44–52.

Gross, M. (1996). *Life on the Edge: Amazing Creatures Thriving in Extreme Environments*. New York: Plenum.

Guilfoile, P. (1997). Wrinkled peas and white-eyed fruit flies: The molecular basis of two classical genetic traits. *American Biology Teacher, 59*, 92–95.

Guilhem, M. (2013, December 15). New princesses rescue girls from a distressed damsel-hood. National Public Radio. Available at http://www.npr.org/2013/12/15/251157298/once-upon-a-time-the-princess-saved-the-environment

Hagen, J. B. (1993). Kettlewell and the peppered moths reconsidered. *BioScene, 19*(3), 3–9.

Hagen, J. B. (1999). Retelling experiments: H. B. D. Kettlewell's studies of industrial melanism in peppered moths. *Biology and Philosophy, 14*, 39–54.

Hagen, J. B., Allchin, D., & Singer, F. (1996). *Doing Biology*. Glenview, IL: Harper Collins.

Hallinan, J. T. (2009). *Why We Make Mistakes*. New York: Broadway Books.

Hamilton, W. D. (1964). The genetical evolution of social behavior. *Journal of Theoretical Biology, 7*, 1–52.

Hanninen, O., Farago, M., & Monos, E. (1983). Ignaz Philipp Semmelweis, the prophet of bacteriology. *Infection Control, 4*(5), 367–370.

Hardwig, J. (1991). The role of trust in knowledge. *Journal of Philosophy, 88*, 693–708.

Harmon, A. (2014, January 5). On Hawaii, a lonely quest for fact. *New York Times*, p. A1. Published online as "A lonely quest for facts on genetically modified crops." Available at http://www.nytimes.com/2014/01/05/us/on-hawaii-a-lonely-quest-for-facts-about-gmos.html

Harmon-Jones, E., & Mills, J. (Eds.). (1999). *Cognitive Dissonance: Progress on a Pivotal Theory in Social Psychology*. Washington, DC: American Psychological Association.

Harris, G. (2011, October 9). Some doctors launch fight against advice on prostate cancer testing. *Star Tribune*, p. A21.

Harvey, W. (1628/1952). *On the Motion of the Heart and the Blood* (R. Willis, Trans.). In *Great Books of the Western World*, Vol. 28. Chicago: Encyclopedia Britannica.

Harvey, W. (1649a/1952). *Anatomical Disquisition on the Circulation of the Blood, to Jean Riolan* (R. Willis, Trans.). In *Great Books of the Western World*, Vol. 28. Chicago: Encyclopedia Britannica.

Harvey, W. (1649b/1952). *A Second Disquisition to Jean Riolan* (R. Willis, Trans.). In *Great Books of the Western World*, Vol. 28. Chicago: Encyclopedia Britannica.

Hauert, C., Traulsen, A., Brandt, H., Nowak, M. A., & Sigmund, K. (2007). Via freedom to coercion: The emergence of costly punishment. *Science, 316*, 1905–1907.

Hauser, M. D. (2006). *Moral Minds*. New York: Ecco.

Henrich, J., McElrath, R., Barr, A., Ensminger, J., Barrett, C., Bolyanatz, A., et al. (2006). Costly punishment across human societies. *Science, 312*, 1767–1770.

Henrich, J., Ensminger, J., McElreath, R., Barr, A., Barrett, C., Bolyanatz, A., et al. (2010). Markets, religion, community size, and the evolution of fairness and punishment. *Science, 327,* 1480–1484.

Herbert, S. (1974). The place of man in the development of Darwin's theory of transmutation, part 1. *Journal of the History of Biology, 7,* 217–258.

Herbert, S. (1977). The place of man in the development of Darwin's theory of transmutation, part 2. *Journal of the History of Biology, 10,* 155–227.

Hertel, C. (2001). Hairy issues: Portraits of Petrus Gonsalus and his family in Archduke Ferdinand II's Kunstkammer and their contexts. *Journal of the History of Collections, 13,* 1–22.

Ho, D. (1999). Bacteriologist Alexander Fleming. *Time 100* [online]. Available at http://www.time.com/time/time100/scientist/profile/fleming.html

Hobbes, T. (1651/1962). *Leviathan*. New York: Collier Books.

Hoekstra, R. F. (1987). The evolution of sexes. In S. C. Sterns (Ed.), *The Evolution of Sex and Its Consequences* (pp. 59–91). Basel: Birkhäuser.

Hofstadter, R. (1955). *Social Darwinism in American Thought* (Rev. ed.). Boston: Beacon Books.

Holmes, D. R. (1989). *Stalking the Academic Communist: Intellectual Freedom and the Firing of Alex Novikoff*. Hanover, VT: University Press of New England.

Holmes, F. L. (2001). *Meselson, Stahl and the Replication of DNA: A History of the Most Beautiful Experiment in Biology*. New Haven, CT: Yale University Press.

Horton, R. (2004). The fool of Pest. *New York Review of Books, 51*(3). Available at http://www.nybooks.com/articles/2004/02/26/the-fool-of-pest/

Howe, E. (2007). Addressing nature-of-science core tenets with the history of science: An example with sickle-cell anemia and malaria. *American Biology Teacher, 69,* 467–472.

Howell, M., & Ford, P. (1980). *The True History of the Elephant Man*. Harmondsworth, UK: Penguin.

Hsü, K. (1986). Darwin's three mistakes. *Geology, 14,* 532–534.

Huber, P. W. (1991). *Galileo's Revenge: Junk Science in the Courtroom*. New York: Basic Books.

Hunter, W. (1784). *Two Introductory Lectures*. London: by order of the Trustees, for J. Johnson.

Hurd, R. (2007). Vitamin C and colds. *National Institutes of Health Medical Encyclopedia*. Available at http://www.nlm.nih.gov/medlineplus/ency/article/002145.htm

Hurford, J. R. (2004). Human uniqueness, learned symbols and recursive thought. *European Review, 12*, 551–565.

Huxley, T. H. (1894/1989). *Evolution and Ethics*. Princeton, NJ: Princeton University Press.

"Ignaz Semmelweis." (n.d.). In *Wikipedia*. Available at http://en.wikipedia.org/wiki/Ignaz_Semmelweis (accessed September 2, 2013).

Jackson, M., & Mendoza, K. (Producers). (1979). Context [video]. In *The Search for Solutions* [video series]. Bartlesville, OK: Phillips Petroleum/Karol Media.

Jensen, K. (2010). Punishment and spite, the dark side of cooperation. *Philosophical Transactions of the Royal Society B, 365*, 2593–2598.

Johnson, S. (2008). *The Invention of Air*. New York: Riverhead Books.

Johnston, W. M. (1972). *The Austrian Mind: An Intellectual and Social History 1848–1938*. Berkeley: University of California Press.

Jones, J. (Director). (1994). Can Chimps Talk? [TV series episode]. *NOVA*, Show No. 2105. Boston: WGBH Educational Foundation. Transcript at http://pin.primate.wisc.edu/aboutp/behavior/nova.html

Judson, H. F. (1979). *The Eighth Day of Creation*. New York: Simon & Schuster.

Judson, H. F. (1981). Chance. In *The Search for Solutions* (pp. 66–85). New York: Holt, Rinehart & Winston.

Kahneman, D. (2011). *Thinking, Fast and Slow*. New York: Farrar, Straus and Giroux.

Katz, L. D. (Ed.). (2000). *Evolutionary Origins of Morality*. Bowling Green, OH: Imprint Academic.

Kelley, T. (2001). *The Art of Innovation: Lessons in Creativity from IDEO, America's Leading Design Firm*. New York: Crown.

Kenner, R. (Director). (2015). *Merchants of Doubt* [Motion picture]. Sony Pictures Classics.

Kettlewell, H. B. D. (1955). Selection experiments on industrial melanism in the Lepidoptera. *Heredity, 9*, 323–342.

Kettlewell, H. B. D. (1956). Further selection experiments on industrial melanism in the Lepidoptera. *Heredity, 10*, 287–301.

Kettlewell, H. B. D. (1959, March). Darwin's missing evidence. *Scientific American 200*(3), 48–53.

Kettlewell, [H.] B. [D.] (1973). *The Evolution of Melanism: The Study of Recurring Necessity; with Special Reference to Industrial Melanism in the Lepidoptera*. Oxford, UK: Clarendon.

Knorr-Cetina, K. (1984). *The Manufacture of Knowledge*. Oxford: Pergamon.

Konnikova, M. (2013, August 8). The psychology of distrusting G.M.O.s. *The New Yorker*. Available at http://www.newyorker.com/tech/elements/the-psychology-of-distrusting-g-m-o-s

Kuhn, T. S. (1970). *The Structure of Scientific Revolutions* (2nd ed.). Chicago: University of Chicago Press.

Langford, D. J., Crager, S. E., Shehzad, Z., Smith, S. B., Sotocinal, S. G., Levenstadt, J. S., Chanda, M. L., Levitin, D. J., & Mogil, J. S, (2006). Social modulation of pain as evidence for empathy in mice. *Science, 312*, 1967–1970.

Latour, B., & Woolgar, S. (1979). *Laboratory Life*. Princeton, NJ: Princeton University Press.

Lawson, A. (2000). The generality of the hypothetico-deductive method: Making scientific thinking explicit. *American Biology Teacher, 62*, 482–495.

Leaf, J. (2011). Charles Keeling & measuring atmospheric CO_2. Minneapolis, MN: SHiPS Resource Center.

Lehrer, J. (2009). *How We Decide*. New York: Houghton Mifflin Harcourt.

Leinhard, J. H. (1997a). Ignaz Philipp Semmelweis. *Engines of Our Ingenuity, No. 622* [Radio series episode]. Houston, TX: KUHF-FM. Available at http://www.uh.edu/engines/epi622.htm (accessed December 30, 2016).

Leinhard, J. H. (1997b). William Harvey. *Engines of Our Ingenuity,* No. 336 [Radio series episode]. Houston, TX: KUHF-FM. Available at http://www.uh.edu/engines/epi336.htm (accessed April 2, 2004).

Lemonick, M. D., & Dorfman, A. (2006, October 9). What makes us different? *Time,* cover, pp. 44–50, 53.

Lenain, T. (1997). *Monkey Painting*. London: Reaktion Books.

Leroi, A. M. (2003). *Mutants: On Genetic Variety and the Human Body*. New York: Viking Adult.

Lewis, R. W. (1988). Biology: A hypothetico-deductive science. *American Biology Teacher, 50*, 362–366.

Lewontin, R. C. (1993). *Biology as Ideology: The Doctrine of DNA*. New York: Harper Collins.

Lewontin, R. C., Rose, S., & Kamin, L. J. (1984). *Not in Our Genes*. New York: Pantheon Books.

Life Magazine. (1996, April 1). Two sisters forever together.

Linden, E. (2003). *The Octopus and the Orangutan*. New York: Plume.

Livio, M. (2013). *Brilliant Blunders: From Darwin to Einstein—Colossal Mistakes by Great Scientists That Changed Our Understanding of Life and the Universe*. New York: Simon & Schuster.

Longino, H. (1990). *Science as Social Knowledge*. Princeton, NJ: Princeton University Press.

Macfarlane, G. (1985). *Alexander Fleming: The Man and the Myth*. Oxford: Oxford University Press.

Magiels, G. (2010). *From Sunlight to Insight: Jan Ingenhousz, the Discovery of Photosynthesis & Science in the Light of Ecology*. Brussels: Academic and Scientific Publishers.

Magner, L. N. (2002). *A History of the Life Sciences* (3rd ed.). New York: Marcel Dekker.

Majerus, M. E. N. (1998). *Melanism: Evolution in Action*. Oxford: Oxford University Press.

Majerus, M. E. N. (2009). Industrial melanism in the peppered moth, *Biston betularia*: An excellent teaching example of Darwinian evolution in action. *Evolution: Education and Outreach, 2*, 63–74.

Malpighi, M. (1661/1929). About the lungs (J. Young, Trans.). *Proceedings of the Royal Society of Medicine, 23*, 1–11.

Malthus, T. (1798/1959). *An Essay on the Principle of Population*. Ann Arbor, MI: Ann Arbor Paperbacks.

Mansfield, H. (2006), *Manliness*. New Haven, CT: Yale University Press.

Martin, B. (1991). *Scientific Knowledge in Controversy*. Albany: State University of New York Press.

Martin, L. J., Hathaway, G., Isbester, K., Mirali, S., Acland, E. L., Niederstrasser, N., Slepian, P. M., Trost, Z., Bartz, J. A., Sapolsky, R. M., Sternberg, W. F., Levitin, D. J., & Mogil, J. S. (2015). Reducing social stress elicits emotional contagion of pain in mouse and human strangers. *Current Biology, 25,* 326–332.

Mason, W.A. (1972). Learning to live. In T. B. Allen (Ed.), *The Marvels of Animal Behavior* (pp. 374–391). Washington, DC: National Geographic.

Masson, J. M., & McCarthy, S. (1995). *When Elephants Weep: The Emotional Lives of Animals*. New York: Delta.

Mathé, J. (1978). *Leonardo da Vinci: Anatomical Drawings*. New York: Miller Graphics.

Matthews, M. (2009). Science and worldviews in the classroom: Joseph Priestley and photo-synthesis. *Science & Education, 18*, 929–960. Reprinted in M. Matthews (Ed.), *Science, Worldviews and Education* (pp. 271–302). Dordrecht, The Netherlands: Springer.

Maurois, A. W. (1959). *The Life of Sir Alexander Fleming, Discoverer of Penicillin*. New York: Dutton.

May, R. (1975). *The Courage to Create*. New York: W. W. Norton.

Mayr, E. (1991). *One Long Argument*. Cambridge, MA: Harvard University Press.

Mayr, E. (1994). Reasons for the failure of theories. *Philosophy of Science, 61*, 529–533.

McCusick, J. (2004). Hypertrichosis universalis congenita, Ambras type; HTC1. *Mendelian Inheritance in Man*. Washington, DC: National Institutes of Health. Available at http://www.omim.org/entry/145701

McGarity, T. O., & Wagner, W. E. (2008). *Bending Science: How Special Interests Corrupt Public Health Research*. Cambridge, MA: Harvard University Press.

McGrew, W.C. (2010). Chimpanzee technology. *Science, 328*, 579–580.

McManus, C. (2002). *Right Hand, Left Hand*. Cambridge, MA: Harvard University Press.

McPherson. (2001). Teaching & learning the scientific method. *American Biology Teacher, 63*(4), 242–245.

Medawar, P. (1964, August 1). Is the scientific report fraudulent? Yes: It misrepresents sci-entific thought. *Saturday Review, 47*, 42–43.

Melis, A. P., & Semmann, D. (2010). How is human cooperation different? *Philosophical Transactions of the Royal Society B, 365*, 2663–2674.

Mendel, G. (1866). Versuche über Pflanzenhybriden [Experiments on plant hybrids]. Reprinted in J. Křiženecký (Ed.), *Fundamenta Genetica* (pp. 57–92). Oosterhout, The Netherlands: Anthropological Publications; Brno, Czechoslovakia: Moravian Museum; & Prague: Czech Academy of Sciences (1965). Eng. trans. reprinted in C. Stern & E. Sherwood (Eds.) (1966), *The Origin of Genetics: A Mendel Source Book* (pp. 1–48). San Francisco: W. H. Freeman. Eng. trans. by C. T. Druery & W. Bateson (1901) avail-able at http://www.mendelweb.org (accessed March 23, 2012).

Mendel, G. (1869/1966). On *Hieracium*-hybrids obtained by artificial fertilization. Eng. trans. reprinted in C. Stern & E. Sherwood (Eds.), *The Origin of Genetics: A Mendel Source Book* (pp. 49–55). San Francisco: W. H. Freeman.

Meselson, M., & Stahl, F. W. (1958). The replication of DNA in *Escherichia coli. Proceedings of the National Academy of Sciences of the USA, 44*, 671–682.

Michaels, D. (2008). *Doubt Is Their Product: How Industry's Assault on Science Threatens Your Health*. Oxford: Oxford University Press.

Milius, S. (2006, July 12). Live prey for dummies: Meerkats coach pups on hunting. *Science News Online*. Available at http://www.sciencenews.org/articles/20060715/fob3.asp

Miller, G. (2005). Spliced gene determines objects of flies' desire. *Science, 308*, 1392.

Miller, G. (2010). The prickly side of oxytocin. *Science, 328*, 1343.

Milloy, S. J. (2001). *Junk Science Judo: Self-Defense against Health Scares and Scams*. Washington, DC: Cato Institute.

Moon, T. J. (1921). *Biology for Beginners*. New York: Henry Holt.

Mooney, C. (2005a). *The Republican War on Science*. New York: Basic Books.

Mooney, C. (2005b). Some like it hot. *Mother Jones, 30*(3), 36–94.

Moore, G. E. (1903). *Principia Ethica*. Cambridge, UK: Cambridge University Press.

Morgan, E. (1990). *The Scars of Evolution*. Oxford: Oxford University Press.

Mulcahy, N. J., & Call, J. (2006). Apes save tools for future use. *Science, 312*, 1038–1040.

Murray, D., Schwartz, J., & Lichter, S. R. (2001). *It Ain't Necessarily So*. Lanham, MD: Rowman & Littlefield.

Murray, R. (2012, August 9). Abby and Brittany Hensel, conjoined 22-year-old twins, get their own reality TV series. *New York Daily News*.

Musch, I., & Willmann, R. (2001). *Albertus Seba's Cabinet of Natural Curiosities*. Koln, Germany: Taschen.

Nash, L. K. (1957). Plants and the atmosphere. In J. B. Conant & L. K. Nash (Eds.), *Harvard Case Histories in Experimental Science* (Vol. 2, pp. 323–435). Cambridge, MA: Harvard University Press.

National Association of Biology Teachers. (2008). The use of animals in biology education [Position statement]. Available at http://www.nabt.org/websites/institution/index.php?p=97

Ness, R. (2012). *Innovation Generation: How to Produce Creative and Useful Scientific Ideas*. New York: Oxford University Press.

New York Academy of Medicine. (2007). A telling of wonders: Teratology in Western medicine through 1800. Available at http://nyam.org/library/collections-and-resources/digital-collections-exhibits/digital-telling-wonders/ (accessed December 30, 2016).

NGSS Lead States. (2013). *Next Generation Science Standards: For States, By States*. Washington, DC: The National Academies Press.

Nickerson, R. S. (1998). Confirmation bias: A ubiquitous phenomenon in many guises. *Review of General Psychology, 2*, 175–220.

Nobel Foundation. (2008). *Nobelprize.org*. Available at http://www.nobelprize.org

Nogler, G. A. (2006). The lesser-known Mendel: His experiments on *Hieracium*. *Genetics, 172*, 1–6.

Novikoff, A. B. (1945a). The concept of integrative levels and biology. *Science, 101*, 209–215.

Novikoff, A. B. (1945b). *Climbing Our Family Tree*. New York: International Publishers.

Novikoff, A. B., & Holtzman, E. (1970). *Cells and Organelles*. New York: Holt, Rinehart & Winston.

Nowak, M. A., Tarnita, C. E., & Wilson, E. O. (2010). The evolution of eusociality. *Nature, 466*, 1057–1062.

Nye, M. J. (2007), Linus Carl Pauling. In *New Dictionary of Scientific Biography* (Vol. 24, pp. 36–44). Detroit, MI: Charles Scribner's Sons.

Online Mendelian Inheritance in Man (OMIM). (2012). *Online Mendelian Inheritance in Man*. Baltimore, MD: McKusick-Nathans Institute for Genetic Medicine, Johns Hopkins University; and Bethesda, MD: National Center for Biotechnology Information, National Library of Medicine. Available at http://omim.org (accessed March 23, 2012).

Orel, V. (1984). *Mendel*. Oxford: Oxford University Press.

Oreskes, N., & Conway, E. M. (2010). *Merchants of Doubt: How a Handful of Scientists Obscured the Truth on Issues from Tobacco Smoke to Global Warming*. New York: Bloomsbury.

Ortony, A. (Ed.). (1993). *Metaphor and Thought*. Cambridge, UK: Cambridge University Press.

Panskepp, J. (2005). Beyond a joke: From animal laughter to human joy? *Science, 308*, 62–63.

Paré, A. (1573/1982). *Of Monsters and Marvels* (J. Pallister, Trans.). Chicago: University of Chicago Press.

Park, R. (2000). *Voodoo Science: The Road from Foolishness to Fraud*. Oxford: Oxford University Press.

Park, R. (2003, January 31). The seven warning signs of bogus science. *Chronicle of Higher Education, 49*, B20.

Parker, J. D. (2004). A major evolutionary transition to more than two sexes? *Trends in Ecology and Evolution, 19*, 83–86.

Pauling, L. (1970). *Vitamin C and the Common Cold*. New York: W. H. Freeman.

Pennisi, E. (2006). Plant wannabes. *Science, 313*, 1229.

Perkins, D. N. (1981). *The Mind's Best Work*. Cambridge, MA: Harvard University Press.

Phillips, R. E., Jr. (2004). The heart and the circulatory system. *Access Excellence Classics Collection*. South San Francisco, CA: Genentech/Access Excellence. Archived at http://accessexcellence.mybiosource.com/files/AE/AEC/CC/heart_background.html

Piaget, J. (1971). *Biology and Knowledge*. Chicago: University of Chicago Press.

Pigliucci, M. (2010). *Nonsense on Stilts: How to Tell Science from Bunk*. Chicago: University of Chicago Press.

Plotkin, H. (1994). *Darwin Machines and the Nature of Knowledge*. Cambridge, MA: Harvard University Press.

Postgate, J. (1994). *The Outer Reaches of Life*. Cambridge, UK: Cambridge University Press.

Potts, R., & Sloan, C. (Curators). (2010a). Hall of human origins [Exhibit]. Washington, DC: Smithsonian Museum of Natural History. Available at http://humanorigins.si.edu

Potts, R. & Sloan, C. (2010b). *What Does It Mean to Be Human?* Washington, DC: National Geographic.

Prebble, J., & Weber, B. (2003). *Wandering in the Gardens of the Mind: Peter Mitchell and the Making of Glynn*. Cambridge, UK: Cambridge University Press.

Priestley, J. (1781). *Experiments and Observations on Different Kinds of Air* (3rd ed.). London: J. Johnson.

Purcell, R. (1997). *Special Cases: Natural Anomalies and Historical Monsters*. San Francisco: Chronicle Books.

Purcell, R. W., & Gould, S. J. (1986). *Illuminations*. New York: W. W. Norton.

Purcell, R. W., & Gould, S. J. (1992). *Finders, Keepers: Eight Collectors*. New York: W. W. Norton.

Quinion, M. (2011). Ahead of the curve. *World Wide Words*. Available at http://www.world-widewords.org/qa/qa-ahe1.htm

Rampton, S., & Stauber, J. (2001). *Trust Us, We're Experts! How Industry Manipulates Science and Gambles with Your Future*. New York: Tarcher/Putnam.

Rand, D. G., Dreber, A., Ellingsen, T., Fudenberg, D., & Nowak, M. A. (2009). Positive interactions promote public cooperation. *Science, 325,* 1272–1275.

Reid, J. B., & Ross, J. J. (2011). Mendel's genes: Toward a full molecular characterization. *Genetics, 189*, 3–11.

Richards, R. (1987). *Darwin and the Emergence of Evolutionary Theories of Mind and Behavior*. Chicago: University of Chicago Press.

Rodgers J. (1991). Mechanisms Mendel never knew. *Mosaic, 22*(3), 2–11.

Rose, S. (1997). *Lifelines: Life beyond the Gene*. Oxford: Oxford University Press.

Rosenberg, S. A. (2011, December 11). Scientists say cod still overfished. *Boston Globe*. Archived at http://archive.boston.com/yourtown/salem/articles/2011/12/11/new_report_could_tighten_cod_fishing_regulations/

Röska-Hardy, L. S., & Neumann-Held, E. M. (Eds.). (2008). *Learning from Animals? Examining the Nature of Human Uniqueness*. London: Psychology Press.

Rottschaefer, W. A. (1998). *The Biology and Psychology of Moral Agency*. Cambridge, UK: Cambridge University Press.

Roughgarden, J. (2004). *Evolution's Rainbow: Diversity, Gender, and Sexuality in Nature and People*. Berkeley: University of California Press.

Rubin, J. (Writer, Producer, Director). (2008). *Ape Genius* [NOVA TV series episode]. Boston: WGBH Educational Foundation.

Rudge, D. W. (1999). Taking the peppered moth with a grain of salt. *Biology and Philosophy, 14,* 9–37.

Rudwick, M. (1974). Darwin and Glen Roy; a "great failure" in scientific method? *Studies in the History and Philosophy of Science, 5,* 97–185.

Rumpf, G. E., & Beekman, E. M. (1999). *The Ambonese Curiosity Cabinet of Georgius Everhardus*. New Haven, CT: Yale University Press.

Ruse, M. (1986). *Taking Darwin Seriously*. New York: Basil Blackwell.

Russell, G. (1996). Biology: The study of life. *AV Magazine, 105*(3), 2–7.

Sapp, J. (1990). The nine lives of Gregor Mendel. In H. E. Le Grand (Ed.), *Experimental Inquiries* (pp. 137–166). Dordrecht, The Netherlands: Kluwer Academic. Also available at http://www.mendelweb.org/MWsapp.intro.html

Schofield, R. (2004). *The Enlightened Joseph Priestley*. University Park: Pennsylvania State University Press.

Semmelweis Society International (2009). Dr. Semmelweis' biography. Available at http://semmelweis.org/about/dr-semmelweis-biography/

Serres, E. R. A. (1832). *Recherches d'Anatomie Transcendante et Pathologique. Theorie des formations et des déformations organiques, appliquée à l'anatomie de Ritta-Christina, et de la duplicité monstrueuse*. Paris: J. B. Balliere.

Shapin, S. (1994). *A Social History of Truth*. Chicago: University of Chicago Press.

Shay, D., & Pinch, T. J. (2005). *Six degrees of reputation: The use and abuse of online review and recommendation systems* (Social Science Research Network, Working Paper Series). Available at http://ssrn.com/abstract=857505 or http://dx.doi.org/10.2139/ssrn.857505

Shepherd, G. M. (2007). John Carew Eccles. In *New Dictionary of Scientific Biography* (Vol. 20, pp. 329–333). Detroit, MI: Charles Scribner's Sons.

Shermer, M. (2002). *Why People Believe Weird Things* (2nd ed.). New York: Holt.

Shubin, N. (2008). *Your Inner Fish*. New York: Random House.

Sigmund, K. (2010). *The Calculus of Selfishness*. Princeton, NJ: Princeton University Press.

Silverstein, A. M. (1989). *A History of Immunology*. San Diego, CA: Academic Press.

Simpson, G. G. (1967). *The Meaning of Evolution* (2nd ed.). New Haven, CT: Yale University Press.

Smith, H. W. (1961). *From Fish to Philosopher*. New York: Anchor.

Smith, J., Van Dyken, D., & Zee, P. C. (2010). A generalization of Hamilton's rule for the evolution of microbial cooperation. *Science, 328,* 1700–1703.

Smith, P. H., & Findlen, P. (Eds.). (2002). *Merchants and Marvels*. New York: Routledge.

Sober, E., & Wilson, D. S. (1998). *Unto Others: The Evolution and Psychology of Unselfish Behavior*. Cambridge, MA: Harvard University Press.

Solomon, M. (2001). *Social Empiricism*. Cambridge, MA: MIT Press.

Spence, W., Herrmann, R. B., Johnston, A. C., & Reagor, G. (1993). *Responses to Iben Browning's prediction of a 1990 New Madrid, Missouri, earthquake* (US Geological Survey Circular 1083). Washington, DC: US Government Printing Office. Available at http://pubs.usgs.gov/circ/1993/1083/report.pdf

Spencer, H. (1851/1969). *Social Statics*. London: John Chapman. Reprinted by Augustus M. Kelley, New York.

Spencer, H. (1852a, March 20). The development hypothesis. *The Leader*. Available at http://www.victorianweb.org/science/science_texts/spencer_dev_hypothesis.html

Spencer, H. (1852b). A theory of population, deduced from the general law of animal fertility. *Westminster Review, 57,* 468–501.

Spencer, H. (1864/1924). *First Principles* (6th ed.). New York: Appleton and Company.

Starr, D. P. (2002). William Harvey. Red Gold: The Epic Story of Blood: Discussion Guide. New York: Educational Broadcasting System.

Stent, G. (Ed.). (1978). *Morality as a Biological Phenomenon*. Berkeley: University of California Press.

Stepan, N. (1982). *The Idea of Race in Science*. London: Macmillan.

Stephens, D. W., McLinn, C. M., & Stevens, J. R. (2002). Discounting and reciprocity in an iterated prisoner's dilemma. *Science, 298,* 2216–2218.

Sulloway, F. (1982). Darwin and his finches: The evolution of a legend. *Studies in the History of Biology, 3,* 23–65.

Sutherland, S. (1992). *Irrationality: Why We Don't Think Straight*. New Brunswick, NJ: Rutgers University Press.

Tanzi, R. E., & Parson, A. B. (2000). *Decoding Darkness*. Cambridge, MA: Perseus.

Tattersall, I. (1998). *Becoming Human: Evolution and Human Uniqueness*. San Diego, CA: Harcourt Brace.

Tattersall, I., & DeSalle, R. (Curators). (2007). What makes us human? Hall of Human Origins [Exhibit]. New York: American Museum of Natural History.

Taylor, J. (2009). *Not a Chimp: The Hunt to Find the Genes That Make Us Human*. Oxford: Oxford University Press.

Thagard, P. (1992). *Conceptual Revolutions*. Princeton, NJ: Princeton University Press.

Thornton, A., & McAuliffe, K. (2006). Teaching in wild meerkats. *Science, 313,* 227–229.

Toumey, C. (1997). *Conjuring Science*. Rutgers, NJ: Rutgers University Press.

Townsley, G. (Writer, Director). (2009). *Becoming Human* [NOVA TV series episodes] (3 Pts.). Boston: WGBH Educational Foundation.

Trivers, R. (1971). The evolution of reciprocal altruism. *Quarterly Review of Biology, 46,* 35–57.

Tschermak, E. (1900/1950). Concerning artificial crossing in *Pisum sativum. Genetics, 35*(5, 2), 42–47.

Twain, M. (1897). *Following the Equator: A Journey around the World*. Hartford, CT: American Publishing Company.

Ule, A., Schram, A., Reidl, A., & Cason, T. N. (2009). Indirect punishment and generosity towards strangers. *Science, 326,* 1701–1704.

"Untangling media messages and public policies." (2012). Understanding Science. Berkeley: University of California Museum of Paleontology. Available at http://undsci.berkeley.edu/article/0_0_0/sciencetoolkit_02

Vogel, G. (2004). The evolution of the golden rule. *Science, 303,* 1128–1130.

Walter, C. (2006). *Thumbs, Toes, and Tears: And Other Traits That Make Us Human.* New York: Walker.

Wang, T. (2002). Peas starch mutants. Norwich, UK: John Innes Center. Available at http://www.jic.bbsrc.ac.uk/STAFF/trevor-wang/appgen/starch/mutants.htm (accessed June 15, 2004).

Warneken, F., & Tomasello, M. (2006). Altruistic helping in human infants and young chimpanzees. *Science, 311,* 1301–1303.

Watson, J. D. (1968). *The Double Helix.* New York: Atheneum.

Weiner, J. (1994). *The Beak of the Finch.* New York: Vintage Books.

Wells, J. (2000). *Icons of Evolution.* Washington, DC: Regnery.

Whitfield, J. (2004, June 6). Everything you always wanted to know about sexes. *PLoS Biology 2,* e183. Available at http://www.pubmedcentral.nih.gov/articlerender.fcgi?artid=423151 (accessed April 10, 2006).

Wilkinson, G. S. (1990, February). Food sharing in vampire bats. *Scientific American, 262*(2), 76–82.

"William Harvey." (n.d.). In *Wikipedia.* Available at http://en.wikipedia.org/wiki/William_Harvey (accessed 2013).

Wimsatt, W. C. (2007). *Re-engineering Philosophy for Limited Beings: Piecewise Approximations to Reality.* Cambridge, MA: Harvard University Press.

Woolfenden, G. E., & Fitzpatrick, J. W. (1978). The inheritance of territory in group breeding birds. *BioScience, 28,* 104–108.

Wright, R. (1994). *The Moral Animal.* New York: Vintage Books.

Yahya, H. [Oktar, A.]. (2006). *The Atlas of Creation,* Vol. 1. Istanbul: Global Publishing. Also available at http://harunyahya.com/ajax/downloadLinks/work/4066

Young, R. (1975). *Darwin's Metaphor: Nature's Place in Victorian Culture.* Cambridge, UK: Cambridge University Press.

Zimmer, C. (2007). *Smithsonian Intimate Guide to Human Origins.* New York: Harper.

IMAGE CREDITS

(1.1) Kunsthistorisches Museum, Vienna. (1.2) Museo Nazionale di Capodimonte, Naples. (2.1) Data and image from Scripps CO_2 Program. (2.2) Courtesy of the Keeling family and the Scripps Oceanographic Institute. (3.1) Photo by Jay Walker, courtesy of Phyllis Novikoff. (3.2) International Publishers, New York. (4.1) Courtesy of Matthew Meselson. (8.1) Photo by Steven Bennett, CC2. (8.2) Photo by David W. Inouye. (9.1) Courtesy of Centre for the Conservation of Specialized Species. (10.1) Courtesy of The Edgar Fahs Smith Memorial Collection, Kislak Center for Special Collections, Rare Books and Manuscripts, University of Pennsylvania. (11.1) Nobel Foundation (2008). (12.1) Darwin (1877, p. 139). (12.2) Watercolor by Conrad Martens (14.1) Photo by Wales Gibbon, CC2. (16.1) Photo by Nick Hobgood, CC2. (16.2) Photo by Roberto Verzo, CC2. (16.3) Darwin (1877, p. 139). (17.1) Serres (1832, plate 1). (18.1a) Doug Cheeseman, Cheeseman Tours. (18.1b) Photo by Maurice Koop, CC2. (18.1c) Photo by Erin Rechsteiner, Hakai Institute. (18.1d) Photo by J. M. Garg, CC3. (18.2) Still from a video by Christophe Boesch, Max Planck Institute for Evolutionary Anthropology. (18.3) Photo by Greg Tee, CC2. (19.1) Photo by Shawn Welling, CC2. (20.1, 20.2) Kettlewell (1973, plate 9.1), Oxford University Press. (21.1) Corbis. (21.3) Wellcome Images of the Wellcome Trust. (22.1) Gregory (1903, p. 226). (23.1, 23.2) Wellcome Images of the Wellcome Trust. (26.1a, left) National Library of Medicine, "Historical Anatomies." (26.1b, right) Gunther von Hagens' BODY WORLDS and the Institute for Plastination, www.bodyworlds.com. (27.1) Creature Comforts Cheshire. (27.2) Courtesy of Charles Poncet, L'Institut National de la Recherche Agronomique.

INDEX